Fuels and Petrochemicals Division 2022

AA002974

Held at the 2022 AIChE Annual Meeting

Phoenix, Arizona, USA
13-18 November 2022

ISBN: 978-1-7138-7896-4

Printed from e-media with permission by:

Curran Associates, Inc.
57 Morehouse Lane
Red Hook, NY 12571

Some format issues inherent in the e-media version may also appear in this print version.

Printed with permission by Curran Associates, Inc. (2023)

For permission requests, please contact AIChE
at the address below.

AIChE
120 Wall Street, FL 23
New York, NY 10005-4020

Phone: (800) 242-4363
Fax: (203) 775-5177

www.aiche.org

Additional copies of this publication are available from:

Curran Associates, Inc.
57 Morehouse Lane
Red Hook, NY 12571 USA
Phone: 845-758-0400
Fax: 845-758-2633
Email: curran@proceedings.com
Web: www.proceedings.com

TABLE OF CONTENTS

ADVANCES IN BIOFUELS PRODUCTION AND ALTERNATIVE FUELS I

25a Thermocatalytic Conversion of Ethanol – A Pathway to Sustainable Aviation Fuel 1
Udishnu Sanyal, Mond F. Guo, Casper O. Brady, Senthil Subramaniam, Karthikeyan Ramasamy

25b Advancement of Catalytic Fast Pyrolysis of Biomass to Produce Renewable Fuels 2
Calvin Mukarakate, Kristiina Ilsa, Huamin Wang, Daniel Santosa, Abhijit Dutta, Joshua Schaidle, Michael B. Griffin

25c Production of Renewable Diesel and Jet Fuel from Thermal Deoxygenation (TDO) Oil 4
Matthew J. Kline, Sampath Karunarathne, Thomas Schwartz, Clayton Wheeler, Hemant P. Pendse

25d Accuracy of Predictions Made by Machine Learned Models for Biocrude Yields Obtained from Hydrothermal Liquefaction of Organic Wastes 5
Feng Cheng, Elizabeth Belden, Wenjing Li, Muntasir Shabuddin, Randy Paffenroth, Michael T. Timko

25e HTL of Wet Wastes for Sustainable Aviation Fuel 7
Michael R. Thorson, Dylan Cronin, Daniel Santosa, Senthil Subramaniam, Andrew J. Schmidt, Karthikeyan Ramasamy, John Norton Jr., Xavi Fonoll Almansa

25f Catalytic Hydrothermal Liquefaction of Food Waste: Improving Product Properties and Increasing Process Economic Viability 8
Hengameh Bayat, Mostafa Dehghanizadeh, Andrea Loya Luján, Catherine Brewer

ADVANCES IN PETROLEUM PRODUCTION AND REFINING

122c Experimental and Numerical Studies on the Phenomena of Interfacial Mixing in Pipelines Processing Lube-Oil 9
Swapana Jerpoth, Robert Hesketh, C. Stewart Slater, Mariano J. Savelski, Kirti Yenkie

122d Preventing H2/bs Poisoning of Pd Membranes for H2/ Purification Using Applied Electric Fields 11
Daniel J. Rivera, Christopher Muhich

122e Understanding Catalyst Failure Mechanisms in Plant Operation 12
Jin Yang, Vera Santos, Ewa Tocha, David G. Barton, Eric Stangland, Daniel Grohol, Van Bloois Stefan, Joost Depicker, Ailene G. Phillips, Michael B Clark, Leadley Stuart, Patrick Heider

122f Strategies for Systematically Optimizing the Operational Management of Lube-Oil Manufacturing and Packaging Facilities 14
Swapana Jerpoth, Robert Hesketh, C. Stewart Slater, Mariano J. Savelski, Kirti Yenkie

122g Demulsification of Tight Crude Oil-Water Emulsions Under Microwave Radiation in Presence of Nanohybrid Hydrophilic Structures 17
Prasad Pawar, Pankti Joshi, Clayton Jeffryes

122h Screening and Investigation on Inhibition of Sediment Formation in a Kuwait Light Crude Oil by Commercial Additives with Some Guidelines for Field Applications ... 18
Ali Qubian, A. S. Abbas, N. Al-Khedhair, Jose Peres, O. Alomair, D. Stratiev, I. Shishkova, M R Riazi

122i Crude Assays - The Distillation ... 26
Andrew Sloley

122j Mechanistic Investigation of Asphaltene-Induced Fouling of Heat Exchangers ... 27
Yousef Al-Shamlan

122k Dual Functional Catalysts for Cracking of Light Cycle Oil into Monoaromatics Using a Hydrogen Donor ... 28
Akshata Ramteke, Dwijraj Mhatre, Divesh Bhatia, Kamal Pant

123a A Co-Axial Microtubular Electrochemical Flow Cell for High-Power Flow Batteries and Electrolyzers ... 30
Nian Liu

DEVELOPMENTS IN ELECTROCHEMICAL REACTORS, FUEL CELLS, AND ELECTROLYZERS II

123b High Performance Direct Methanol Fuel Cells Incorporating pH-Gradient-Enabled Microscale Bipolar Interfaces ... 31
Kritika Sharma, Shrihari Sankarasubramanian, Zhongyang Wang, Vijay Ramani

123c Intensifying Thermocatalytic Methanol Synthesis Using Electrochemical H_2 Activation ... 33
Samantha Roenigk, Evan Miu, Manjodh Kaur, James R. McKone

123d Separation of Diluted Ethanol Produced by CO_2 Electrolysis Can Be Cost Effective ... 34
Magda Barecka, Joel W. Ager III, Alexei A. Lapkin

123e Model-Based Development of a Novel Apparatus for Electrochemically Induced Crystallization of (di-)Carboxylic Acids ... 35
Jonas Goertz, Kristina Mielke, Marcel Gausmann, Andreas Jupke

123f Investigation of Current Production and Metal Removal from Synthetic Desalter Effluent Using a Dual Chamber Microbial Fuel Cell (MFC) with <em>Shewanella Oneidensis</Em> MR-1 ... 37
Carlos Munoz, Amarjeet Bassi

635a First European Demonstration Plant for the Synthesis of Renewable Ome Fuels ... 47
Alvaro Ferre, Johannes Voggenreiter, Christian F. Breitkreuz, Denis Worch, Udo Lubenau, Hans Hasse, Jakob Burger

DEVELOPMENTS IN ALTERNATIVE FUELS AND ENABLING TECHNOLOGIES

635b Additive Solutions for Renewable Fuels and Feedstocks ... 50
Daniel Dreyer, Nestor Soriano Jr., Ashish Dhawan, Oussama Zenasni, Josep Marca, Elizabeth Balapitiya

635c CFD Modeling Development of Onboard Absorptive Hydrogen Storage on Coal-Derived Activated Carbon for Light-Duty Vehicles ... 58
Subhodeep Banerjee, Ki-Joong Kim, Christopher Matranga, Mehrdad Shahnam, William Rogers

635f Investigating the Performance of the Pyrolysis Process for Different Feedstocks 59
Hasan Al-Abedi, Haider Al-Rubaye, Joseph Smith

635g Total Acid Number Reduction of Renewable Thermal Deoxygenation (TDO) Oil for Coprocessing in Refineries 60
Sampath Karunarathne, Matthew J. Kline, M. Clayton Wheeler

65a Carbon-Free Platinum Bismuth Alloy Nanoplatelets with Enhanced Activity and Platinum Stability in Operating Proton Exchange Membrane Fuel Cells and Electrolyzers 61
Junchuan Fang, Pranjul Mani Dubey, Anastasios Angelopoulos

DEVELOPMENTS IN ELECTROCHEMICAL REACTORS, FUEL CELLS, AND ELECTROLYZERS I

65b Proton-Conducting Solid Oxide Electrolysis Cells Based on Doped-$BaZrO_3$ 62
Clarita Regalado Vera, Hanping Ding, Jagoda M. Urban-Klaehn, Meng Zhou, Hongmei Luo, Dong Ding

65c Design and Development of a Kilowatt Scale Stack for Carbon Dioxide and Carbon Monoxide Electroreduction 63
Sean Overa, Feng Jiao

65d Modelling the Electrochemical Reduction of CO_2 Under Gas-Liquid Taylor Flow 64
Isabell Bagemihl, Chaitanya Bhatraju, J Ruud Van Ommen, Volkert Van Steijn

65e Modular Dynamic Modeling of Electrochemical Reactors 65
Dominik Bongartz, Luisa Carola Brée, Katharina Maria Ebeling, Matthias Leitl, Georgia Ioanna Prokopou, Jan Raphael Seidenberg, Alexander Mitsos

65f Tailoring Electrode Microstructures for High-Performance Redox Flow Batteries Through Non-Solvent Induced Phase Separation 66
Charles Wan, Rémy Jacquemond, Antoni Forner-Cuenca, Yet-Ming Chiang, Fikile R. Brushett

256a Reservoir Quality Indicators and Relative Thermal Maturity Classification Using Image-Based and Bulk Rock Characterization 68
Shannon Eichmann, David Jacobi, Poorna Srinivasan, Jennifer Rodriguez

DEVELOPMENTS IN SHALE GAS AND NATURAL GAS

256b Benefits of Removing Nitrogen Before the Cold Section in Baseload LNG Plant 69
Ajinkya Pal, Iftekhar Karimi, Easa Al-Musleh

256c Modeling Uncertainty and Impact of Non-Darcy Flow Regimes on Unconventional Well Performance 71
Harun Ates, Shannon Eichmann, Kanhaiyalal Patel, Ravi Vaidya, Rabah Mesdour

256d Elucidation of Non-Catalytic Ethylene Polymerization Reactions Through Computational Study 72
Alexander Shaw, Grant Marsden, Linda Broadbelt

256e Functionalized Iron Oxide (Fe_3O_4) Nanoparticles as a New Type of Recyclable Hydrate Dispersant 73
Yuanxing Zhang, Ning Wu, Amadeu Sum

319a Hydrogen Generation from Natural Gas Using Autothermal Chemical Looping Systems: Process Simulation and Heat Integration of Two Process Configurations........ 74
Fanhe Kong, Yitao Zhang, Pinak Mohapatra, Andrew Tong, Liang-Shih Fan

FUEL PROCESSING FOR HYDROGEN PRODUCTION

319b Catalytic Conversion of Methane to Hydrogen and Carbon Nanotubes by Microwave Irradiation........ 75
Changle Jiang, Brandon Robinson, Sonit Balyan, Jianli Hu

319c Exploring Production Aspects of Blue Hydrogen and Valuable CNTs from Natural Gas in a Fluidized Bed Reactor........ 76
Kaushal Parmar, Kamal Pant, Shantanu Roy

319d Compact Flex-Fuel Processors for Syngas and Hydrogen Production........ 77
Hani Hawa, Saurabh A. Vilekar, Christian Junaedi, Subir Roychoudhury

319e Ni-MgO Based NOVEL Catalysts Prepared by Combustion Synthesisfor CO_2 and Methane Reforming........ 78
Vardan Danghyan, Alexander Mukasyan, Anand Kumar, Eduardo E. Wolf

319f Solution Combustion Synthesis of Iron-Based Alumina Nanocomposites for Microwave-Assisted Thermocatalytic Dehydrogenation of Fossil Fuels........ 79
Zachary Chanoi, Victoria Reyes, Evgeny Shafirovich

319h Exploring the Potential of Ni_3S_2 Based Sulfur Looping Process for Efficient Production of H_2 from H_2s: A Thermodynamic and Kinetic Study........ 80
Anuj Joshi, Kalyani Jangam, Zain Mohammad, Liang-Shih Fan

319i Integrated SOLID OXIDE Electrolytic Cell for a NOVEL Hydrogen Production Using Water-Energy Nexus Framework........ 81
Lateef Jolaoso, Javad Asadi, Pejman Kazempoor

535a Acetylene Powered Thermal Power Plant Boiler........ 84
Debshilpi Patra

POSTER SESSION: FUELS AND PETROCHEMICALS DIVISION

535b A Novel Fluid System to Integrate Stimulation, EOR, and CO2 Storage in Shale Oil Reservoirs........ 85
Yang Zhao, Yanxu Ding

535c Fluoro- Vs Hydrocarbon Surfactants: Synthesis and Application in Enhanced Oil Recovery........ 86
Syed Hussain, Muhammad Shahzad Kamal, Ahmad Mahboob, Mobeen Murtaza, Shirish Patil, Emad W. Al Shalabi

535d Fast Pyrolysis of Food Waste in N_2 and CO_2 After Torrefacton in a Bubbling Fluidized Bed Reactor........ 87
Seung-Soo Kim, Hoang Vu Ly, Jinsoo Kim, Hyun Tae Hwang

535e Thermogravimetric and Kinetic Study of Hemp and Milkweed........ 88
Seung-Soo Kim, Jinsoo Kim, Foster Agblevor

535f pH and Salt-Responsive Dynamic Binary Complexes Based on Supramolecular Complexation with Applications in Unconventional Reservoirs 89
Bhargavi Bhat, Shuhao Liu, Yu-Ting Lin, Joseph Kwon, Mustafa Akbulut

535g Upgrading Tire-Derived Pyrolysis Oil Via Distillation for Blending with Fuel Oils 90
Ryan Dantu, Johannes Knoetze, Cara Schwarz

535h Development of Bioethanol from Molasses: A Modern Approach 93
Ayush Gondane, Tanashree Paraye, Utkarsh Maheshwari

535i Crude Assays Cut Properties 94
Andrew Sloley

535j Screening Criteria and Experimental Investigation for Enhanced Oil Recovery at Umm Niqa Oil Field-Kuwait 95
Ahmed Alshammri, Osamah Alomair, Adel Elsharkawy

395a Polyazeotropy for Binary Mixtures: Methanol and Diethylamine 96
Nooram Anjum, Shigeo Oba, Toshihiko Hiaki, Chau-Chyun Chen

PROPERTIES AND PHASE EQUILIBRIA FOR FUELS AND PETROCHEMICALS

395b the Investigation of the Effects of Asphaltene Molecular Properties on Asphaltene-Solids Heterogeneous Interaction 97
Weiyi Kong, Rizwanur Rahman, Michael P. Hoepfner

395c the Ideal Gas Contribution to Liquid Heat Capacity 98
Cassandra Guffey, Joseph Bloxham, Neil Giles, Thomas Knotts Iv, Wade Vincent Wilding

395d Modeling of Nitrogen + Fuel Phase Behavior Using the Helmholtz-Energy-Explicit Equation of State 99
Aaron Rowane, Ian H. Bell

395e Improving the Capability of Process Simulators to Represent Solid-Fluid Equilibria Applied to Natural Gas Liquefaction: The Methane + Neopentane System at Low Temperature 100
David Bluck, Seiya Hirohama, Nevin Gerek Ince, Marco Campestrini, Paolo Stringari, Freddy Garcia, Jean-Jacques Bartuel, Gang Xu

395f Methane Mass Transfer in Liquid Hydrocarbon Mixtures – Measurement and Modeling in Synthetic Porous Media 101
Sheng Hu, Theodore Tsotsis, Kristian Jessen

395g Exploring Liquid Crystalline Phase Behavior of Pyrene Pitch Oligomers Isolated Via Supercritical Extraction 102
Graham W. Tindall, Jefferson T. Giles II, Adam S. Gross, Yunlong Zhang, Amy C. Clingenpeel, Kazem V. Edmond, Manesh Gopinadhan, Stuart E. Smith, David Perkins, Jonathan D. Saathoff, Mark C. Thies

599a Solutions of Nonionic Surfactants in CO2 to Improve Unconventional CO_2-EOR Via Wettability Alteration 103
Lauren Burrows, Foad Haeri, Deepak Tapriyal, Sean Sanguinito, Parth Shah, Peter Lemaire III, Robert M. Enick, Dustin Crandall, Angela Goodman

UNCONVENTIONAL OIL AND NATURAL GAS: SCIENCE AND TECHNOLOGY ADVANCEMENT

599e Production Performance of US Oil Shale Plays Related to Their Depth Profile and Initial Gas Oil Ratio (GOR) 104
Rasoul Sorkhabi, Palash Panja

599f A Study of Capillary Condensation Phenomena in Mesoporous Materials 105
Jiyue Wu, Sheng Hu, Kristian Jessen, Theodore Tsotsis

396a Towards Rational Design of Metal-Embedded Zeolite Catalyst for Improved Catalytic Performance in Microwave-Assisted Methane Dehydro-Aromatization 106
Sanjana Karpe, Xinwei Bai, Brandon Robinson, Jianli Hu, Goetz Veser

VALUE-ADDED CHEMICALS FROM NATURAL GAS I

396b Deactivation Study of Mo-HZSM-5 Catalyst for Microwave-Assisted Methane Aromatization: Effect of Reaction Parameters 108
Ashraf Abedin, Pranjali Muley, Victor Abdelsayed, Xinwei Bai, Hari Paudel, Daniel Haynes

396c Ethane Dehydroaromatization with CO_2 Co-Feed Over Microwave Synthesized Core-Shell Transition-Metal-Carbide@ZSM-5 Material 110
Ashley Caiola, Brandon Robinson, Sean Brown, Jianli Hu

396d Enhanced Dry Reforming of Methane Using Pseudo Catalytic Metal Oxide and Nanoparticles 111
Pinak Mohapatra, Qichang Meng, Anuj Joshi, Sonu Kumar, Patricia Loughney, Ashin Sunny, Lang Qin, Zhuo Cheng, Vicky Doan-Nguyen, Liang-Shih Fan

396e Low Temperature "Super-Equilibrium" Reforming of Methane Through Chemical Looping 113
Luke Neal, Chongyan Ruan, Fanxing Li

396f Microwave Catalytic Conversion of Acetylene for Co-Production of Hydrogen and Carbon Nanotubes 114
Sonit Balyan, Changle Jiang, Ashley Caiola, Jianli Hu

195a Building a Toolbox for Integrating Economic and Environmental Performance of Bio-Based Products 115
Samir Meramo, Peter Fantke, Sumesh Sukumara

INTEGRATED BIOREFINERIES: TECHNOLOGIES AND TEA/LCA

195b Enhanced Sugar Yield from Cellulose by Ammonium Salt Based Pretreatment 117
Monidipa Bose, Eshita Saha, Shyamal Roy

195c TEA and LCA of Renewable Hydrocarbon Fuels and Co-Products from Lignocellulosic Biomass Via 2,3 Bdo Fermentation, Separation, Dehydration and Oligomerization. 118
Marco Avendano, Qiang Fu, Jianpei Lao, Oindrila Gupta, Mi Lu, Sankar Nair, Matthew Realff

195d Advanced Biorefinery Feedstocks from Waste Biomass: Technoeconomic Assessment of Integrated Air Classification-Hydrothermal Carbonization-Pelletization Processes 120
Md Tahmid Islam, Nepu Saha, Jordan Klinger, Toufiq Reza

195f Oxidative Fast Pyrolysis of Groundnut Shell (GNS) in the Catalytic and Non-Catalytic Mode for Biofuel Production 121
Geeta Kumari, Bhavin Soni

195g TEA/LCA of Integrated Biochemical and Hydrothermal Processing of Corn Stover for Fuels and High Value Products 122
Bharathkiran Maddipudi, Khang Huynh, Vinod S. Amar, Katelyn Shell, Anuj Thakkar, Anuradha Shende, Sergio Hernandez, John E. Aston, Sandeep Kumar, Ram Gupta, Rajesh Shende

64a Evaluation of Combustion Properties of Corn Stover-Derived Solid and Liquid Biofuels by Cone Calorimeter 123
Md Tahmid Islam, Jordan Klinger, Toufiq Reza

ADVANCES IN BIOFUELS PRODUCTION AND ALTERNATIVE FUELS II

64c Production of Sustainable Aviation Fuels from Lignocellulosic Biomass Feedstocks 124
Foster Agblevor, Hossein Jahromi

64e Evaluation of Ultra-High Temperature Resistant Preformed Particle Gels for Geothermal Preferential Fluid Flow Control 125
Tao Song, Thomas Schuman, Baojun Bai

64f Roles of Urea Inclusion Fractionation of Biodiesel on the Cold Flow Properties and Establishment of Feedstocks for Biochemicals and Biomaterials 126
Junli Liu, Bernard Tao

64g Novel Supercritical Biodiesel Plant Design and Process Scale-Up 127
Shyam Paudel, Caleb Moellenhoff, Haider Al-Rubaye, Joseph Smith

453a Direct Methane Conversion by Atomically Thin Platinum Nanolayers on Two Dimensional Metal Carbides 128
Yue Wu

VALUE-ADDED CHEMICALS FROM NATURAL GAS II

453b Oxidative Coupling of Methane Catalysts: Developing Structure-Property Relationships for High-Performance Metal Oxide Systems 129
Mariano D. Susman, Matteo Ferri, Raffaele Cheula, Hien N. Pham, Abhaya K. Datye, Sivadinarayana Chinta, David West, Matteo Maestri, Jeffrey Rimer

453c Pt-Based Monolithic Catalysts for Oxidative Coupling of Methane: Effect of Catalyst Formulation and Operation Parameters 130
Sven Schardt, Ahmet Celik, Jaspreet Chawla, Simon Bastian, Stephan A. Schunk, Patrick Lott, Olaf Deutschmann

453d New Insights into the Effect of CO2 on Oxidative Coupling of Methane (OCM) 133
Yonggang Cheng, Pedro Mendes, Joris Thybaut

453e Non-Catalytic Direct Partial Oxidation of Methane to Methanol in a Microreactor 136
Kelly Cohen, Justin Blanchard, James Dorman, Kerry Dooley

453f Electrochemical Oxidation of Methane to Methanol on Electrodeposited Transition Metal Oxides Under Ambient Temperature 137
Kangze Shen, Yu-Chao Huang, Joonbaek Jang, Carlos Morales-Guio

453g Ethane Dehydrogenation Using Liquid Metal Catalysts 139
Aime Laurent Twizerimana, Kaylen Ocampo, Aaditya Hari Bharanidharan, Kayle McGaughy, Mohammad S. Masnadi, Goetz Veser

453h Ethane Oxidative Dehydrogenation by CO_2 Over Stable Csru/CeO_2 Catalyst 140
Xiaoyan Wang, Yuxin Wang, Brandon Robinson, Jianli Hu

Author Index

25a Thermocatalytic Conversion of Ethanol – a Pathway to Sustainable Aviation Fuel

Udishnu Sanyal, Mond F. Guo, Casper O. Brady, Senthil Subramaniam, Karthikeyan Ramasamy

Commercial aviation fuel is responsible for ~13% of transportation greenhouse gas (GHG) emission. To meet the target of complete decarbonization by 2050, airlines have committed to carbon-neutral growth which requires non-fossil sourced fuels or sustainable aviation fuels (SAF). While electrification is not viable, decarbonization of SAF requires an effort to develop new economically viable technologies from biobased or waste carbon resources. In this direction, ethanol, largest biobased product manufactured in the world shows tremendous potential due to its availability in large quantities and at reasonable price. Global production of ethanol in the United States reaches about 16 billion gallons/year in various distributed facilities. Given the increasing availability of ethanol, building additional markets via the novel ethanol processing technologies that are robust and economically viable is a foreseeable pathway for creating value to ethanol producer.

In the present study we developed a process that utilizes wet ethanol as the feedstock and convert it to SAF in a series of chemical steps. The first step of this process is ketonization of the ethanol which generated acetone, 2-pentanone and 2-heptanone. Over the different mixed metal oxide catalyst, Pd promoted ZnO-ZrO2 exhibited the highest performance. *In-situ* formation of Pd-Zn alloy was attributed to the active site which promoted rapid dehydrogenation and hydrogen transfer in conjunction of efficient C-C coupling on acid-base sites that form long chain ketones.[1] In the next step, when mixed ketones were subjected to Pd/Mg_4AlO catalyst a pool of C_8-C_{18} ketones were formed via aldol and cross aldol condensation which upon hydrogenation produces corresponding alkanes and iso-alkanes which are ideal for jet fuels. When Mg_4AlO was obtained as catalyst cyclic ketones were obtained as the products which typically falls beyond the jet range (>C_{15}). Additionally rapid deactivation of the catalysts was also noted in this case. Introducing small amount of Pd as promoter not only prevents the catalyst deactivation but also generates mixture of oxygenates that are required for jet range (C_9-C_{16}). Effect of different process parameters was investigated which shows an effect on products obtained. The difference in reactivity was attributed to difference in active sites and overall reaction mechanism which will be discussed in detail.

1. Subramaniam, S.; Guo, M. F.; Bathena, T.; Gray, M.; Zhang, X.; Martinez, A.; Kovarik, L.; Goulas, K. A.; Ramasamy, K. K. *Angew Chem. Int. Ed.* **2020**, *59*, 2-10.

25b Advancement of Catalytic Fast Pyrolysis of Biomass to Produce Renewable Fuels

Calvin Mukarakate, Kristiina Iisa, Huamin Wang, Daniel Santosa, Abhijit Dutta, Joshua Schaidle, Michael B. Griffin

Catalytic fast pyrolysis (CFP) using bifunctional metal-acid catalysts is a practical approach for converting biomass into high yields of stabilized biocrude (CFP-oil) for either standalone hydrotreating or coprocessing in existing petroleum refineries to produce fuels. We have been developing an integrated fixed-bed CFP technology coupled with hydrotreating (HT) to produce hydrocarbon fuel blendstocks from lignocellulosic biomass[1-2]. Experiments conducted in a bench-scale reactor using Pt/TiO_2 with co-fed H_2 demonstrated production of CFP-oils with ≥35% carbon yields and <20wt% oxygen contents on dry basis. The CFP process and Pt/TiO_2 catalyst robustness was demonstrated by evaluating 50% clean pine and 50% forest residue blends and over 100 cycles with the same catalyst bed. Hypothesis-driven modification to the Pt/TiO_2 led to greater than 500% improvement in time-on-stream (2h vs > 15h) per CFP cycle while producing oils with similar carbon yields and oxygen contents. The residual oxygen in the Pt/TiO_2 CFP-oils was reduced to <1wt% using a single-stage standalone hydrotreater that was operated for over 100h to produce hydrocarbon blendstocks. The integrated CFP—HT results were coupled with technoeconomic analysis and achieved a modelled minimum fuel selling price (MFSP) of $3.33/GGE, assuming capture and recovery of coproducts generated during the CFP process. Even though these results were encouraging there are still important research questions that need to be addressed to make the integrated CFP—HT technology cost competitive and to scale up. Recently we demonstrated that capturing CFP-derived coproducts and co-hydrotreating of the Pt/TiO_2 CFP-oil with straight-run diesel can lower MFSP to <$3.00/GGE and reduce GHG emissions by 60% compared to petroleum derived gasoline. This presentation seeks to demonstrate progress made in advancing the integrated CFP—HT process by linking catalyst performance and process development, guided by technoeconomic and life-cycle analyses, and highlighting the importance of computational modeling[3-4] in scaling up the technology.

References

1. Griffin, M. B.; Iisa, K.; Wang, H.; Dutta, A.; Orton, K. A.; French, R. J.; Santosa, D. M.; Wilson, N.; Christensen, E.; Nash, C.; Van Allsburg, K. M.; Baddour, F. G.; Ruddy, D. A.; Tan, E. C. D.; Cai, H.; Mukarakate, C.; Schaidle, J. A., Driving towards cost-competitive biofuels through catalytic fast pyrolysis by rethinking catalyst selection and reactor configuration. *Energy andamp; Environmental Science* **2018,** *11* (10), 2904-2918.
2. French, R. J.; Iisa, K.; Orton, K. A.; Griffin, M. B.; Christensen, E.; Black, S.; Brown, K.; Palmer, S. E.; Schaidle, J. A.; Mukarakate, C.; Foust, T. D., Optimizing Process Conditions during Catalytic Fast Pyrolysis of Pine with Pt/TiO2—Improving the Viability of a Multiple-Fixed-Bed Configuration. *ACS Sustainable Chemistry andamp; Engineering* **2021,** *9* (3), 1235-1245.
3. Pecha, M. B.; Iisa, K.; Griffin, M.; Mukarakate, C.; French, R.; Adkins, B.; Bharadwaj, V. S.; Crowley, M.; Foust, T. D.; Schaidle, J. A.; Ciesielski, P. N., Ex situ upgrading of pyrolysis vapors over PtTiO2: extraction of apparent kinetics via hierarchical transport modeling. *Reaction Chemistry andamp; Engineering* **2021,** *6* (1), 125-137.

4. Adkins, B. D.; Mills, Z.; Parks Ii, J.; Pecha, M. B.; Ciesielski, P. N.; Iisa, K.; Mukarakate, C.; Robichaud, D. J.; Smith, K.; Gaston, K.; Griffin, M. B.; Schaidle, J. A., Predicting thermal excursions during in situ oxidative regeneration of packed bed catalytic fast pyrolysis catalyst. *Reaction Chemistry andamp; Engineering* **2021,** *6* (5), 888-904.

25c Production of Renewable Diesel and Jet Fuel from Thermal Deoxygenation (TDO) Oil

Matthew J. Kline, Sampath Karunarathne, Thomas Schwartz, Clayton Wheeler, Hemant P. Pendse

Demand for renewable fuels is projected to increase, and one of the most promising materials to convert into bio-oils is cellulose. Residue from logging, farming, or even municipal waste can be utilized as a potential feedstock. Several processes have been developed that convert cellulose into bio-oil, one of them being Thermal DeOxygenation (TDO). It is a process that converts organic acids from cellulose hydrolysis and dehydration into a low-oxygen bio-oil (< 5 wt %) with a broad boiling point distribution (75-550 °C). Additionally, the compounds can be upgraded into diesel and jet fuel fractions. However, catalytic hydrotreating and ring-opening reactions are required for TDO oil to meet the specifications for middle distillate fuels.

In this work, a mixture of levulinate and formate salts was converted to crude TDO oil in a pilot-scale continuous flow reactor at 450 °C and atmospheric pressure. Using a high-surface-area Ni catalyst (Ni/SiO_2-Al_2O_3), TDO oil was hydroprocessed to saturate aromatic rings and convert the oxygenated compounds in the fuel. This hydrotreated material was shown to have significantly better combustion characteristics, but improvements in the cetane number, hydrogen content, and density were needed to meet the requirements for No. 2 distillates and jet fuels.

For the ring-opening (endocyclic C-C hydrogenolysis) of hydrotreated TDO oil, several supported Ir catalysts were evaluated to produce high-cetane blendstocks. Ir catalysts with high activity and selectivity to desired products were chosen as candidates for scale-up reactions and the fuel was characterized and compared to fuel specifications. Using the catalysts presented in this research, we are able to produce renewable diesel and jet fuels that can be used as blending agents or as drop-in fuels.

25d Accuracy of Predictions Made By Machine Learned Models for Biocrude Yields Obtained from Hydrothermal Liquefaction of Organic Wastes

Feng Cheng, Elizabeth Belden, Wenjing Li, Muntasir Shabuddin, Randy Paffenroth, Michael T. Timko

Hydrothermal liquefaction (HTL) has potential for converting the world's massive quantities of wet organic wastes into renewable fuels. Wet wastes fed to HTL undergo a complex series of reactions, resulting in formation of an energy-dense biocrude product. The biocrude yield is one of the most important parameters determining economic viability, making its prediction for a given feedstock an important capability for allocating finite resources for investment in HTL commercialization. The complex reaction network, consisting of hundreds of reactants and thousands of reactions, is a challenge for traditional, physics-based predictions of reaction outcomes. Data-driven techniques constitute a newly emerging alternative to physics-based prediction methods. Ensuring the accuracy of data-driven models requires careful data curation, model selection, and method development. In this work, a data set was assembled consisting of 570 data points that had previously been published in the open literature After dividing the data set into training and testing sub-sets, eight different algorithms were tested for accuracy, using seven different compositional properties as the independent variables. Among the tested algorithms, Random Forest and eXtreme Gradient Boosting (XGBoost) made the most accurate predictions of biocrude yield compared with the test set, with root mean square errors (RMSE) of 8.34 and 8.57, respectively. Further refinement of the Random Forest model reduced its RMSE to 8.07. Predictions made by a series of literature models for the same test set data resulted in RMSE ranging from 9.2 in the most accurate case to 27.6. Finally, eight feeds that have never been studied previously using HTL were selected from the literature and biocrude yields predictions were made for them. The biocrude yields and corresponding uncertainties were fed into a probabilistic economic model to project the minimum fuel selling price for a break-even process. Economic projections clearly showed that the machine learning

regression can differentiate between economically viable and non-viable feeds and, due to the relationship between biocrude yield and economic performance, can even differentiate between economically viable feeds with discernment. When combined with other parameters, including especially process scale and feedstock cost, the new machine learned regression is a valuable tool for guiding investment of finite resources.

25e HTL of Wet Wastes for Sustainable Aviation Fuel

Michael R. Thorson, Dylan Cronin, Daniel Santosa, Senthil Subramaniam, Andrew J. Schmidt, Karthikeyan Ramasamy, John Norton Jr., Xavi Fonoll Almansa

Hydrothermal liquefaction (HTL) uses heat and pressure to liquefy the organic matter in biomass/waste feedstocks to produce biocrude. Hydrotreating converts the biocrude to transportation fuels, including SAF. This presentation will investigate the potential for utilizing wet-waste HTL biocrude as an SAF feedstock. Herein, several biocrude samples were produced from food waste, sewage sludge, and fats, oils, and grease, and subsequently hydrotreated and distilled to produce SAF samples. The fuels (both undistilled and distilled samples) were analyzed via elemental and 2D-GC-MS. Herein, we report the Tier α and β analysis of an SAF sample derived originally from a biocrude. The upgraded biocrude in the SAF ranges exhibits key fuel properties, including carbon number distribution, distillation profile, surface tension, density, viscosity, heat of combustion, and flash point, which all fall within the required range for aviation fuel. This presentation will cover both the attractive features of SAF from HTL as well as the remaining technical challenges.

25f Catalytic Hydrothermal Liquefaction of Food Waste: Improving Product Properties and Increasing Process Economic Viability

Hengameh Bayat, Mostafa Dehghanizadeh, Andrea Loya Luján, Catherine Brewer

Hydrothermal liquefaction (HTL) is a promising thermochemical conversion technology for biomass valorization and bio-crude oil production. Moving this technology towards commercialization still faces several challenges. Enhancing the economic feasibility and sustainability of the HTL process requires improvement of the bio-crude oil properties and effective valorization of the HTL co-products (aqueous phase and char) for fuels, nutrients and materials.

This study investigated the HTL of food waste using red mud as a catalyst with the goals of increasing the bio-crude oil yield and improving the physicochemical properties of the bio-crude oil and the HTL-char. HTL of food waste (15 wt.% solids loading) and red mud (5 wt.%) was conducted using a 100 ml batch reactor at 240-290 °C with 0-60 min reaction times.

Preliminary results showed that bio-crude yield from HTL at 240 °C and 30 min increased from 27.8 wt.% to 35 wt.% when red mud was added; the bio-crude oil higher heating value showed a slight increase from 36.1 to 37.5 MJ/kg. Bio-crude oil characterization revealed similar families of molecules with and without the presence of the catalyst.

Food waste HTL-char and catalytic HTL-char were tested adsorbents for the removal of lead, copper, and nitrate from wastewater. Batch experiments were used to study the effects of char modification, initial wastewater pH, adsorption contact time, and initial metal concentration on adsorption performance. Surface characterization of HTL-chars using X-ray diffraction (XRD), Fourier transform infrared (FTIR), and N_2 gas adsorption analysis indicated the red mud particles were successfully introduced into the HTL-char during HTL.

122c Experimental and Numerical Studies on the Phenomena of Interfacial Mixing in Pipelines Processing Lube-Oil

Swapana Jerpoth, Robert Hesketh, C. Stewart Slater, Mariano J. Savelski, Kirti Yenkie

Pipelines are considered the most crucial mode of transportation in the chemical and petroleum industry. While a single-product pipeline system is comparatively easier to manage, The multi-product pipeline system poses many difficulties due to the high level of operational complexities (Kirschstein 2018). In the multi-product pipeline systems, different types and grades of products are successively transferred in batches. In these pipelines, a major challenge that has been long-standing and economically significant is the interfacial mixing of consecutive product batches when switching from one product to another. The changeover operation leads to the generation of contaminated/ mixed products at the interface of adjacent batches. The commingled product does not match the desired specifications of either of the two batches and is categorized as a downgraded product, which cannot be used for the desired application. To address this challenge, over the years, extensive research has been carried out, and systematic scheduling plans have been developed by industries (Liao et al. 2019; Dimas et al. 2018). Through these plans, the volume of contamination is reduced by preventing the products that have significant differences in physical properties from being processed adjacently. However, these scheduling plans are not completely efficient and therefore the industries undergo high economic losses due to the generation of large volumes of mixed products. Hence, another way to address this problem is to understand the interfacial mixing by studying the transport properties of the fluids (Patrachari and Johannes 2012). Over the years, several mathematical models have been reported by researchers for estimating the volume of the mixed fluid (Ramanujan 2008). The reported mathematical models are primarily empirical in nature and case study specific. To this end, our work focuses on the case study of a lube-oil processing pipeline that manufactures and packages over a thousand different products throughout the year.

Typically, to avoid contamination due to mixing, the following operations are adopted by a generic lube-oil industry, the pipelines are first gravity drained and then blown using compressed air. However, due to the high viscosities of the lube oil, gravity draining and compressed air blowing are not sufficient for efficiently cleaning the pipelines and therefore residues of the prior oil remain adhered to the inner pipeline walls. Hence, followed by air-blowing the pipelines are flushed by using the lube oil from the batch that is processed next. During the flushing operation, the residual oil from the pipeline walls diffuses into the fresh batch of oil and results in contamination. The flushing is continued until the new batch of oil meets the desired specifications when tested for various physical properties in the laboratory. The existing operation often results in excessive flushing, is based on trial and error, and leads to improper resource consumption and energy conservation. Hence, our work aims to help the lube-oil industries make informed decisions and minimize the generation of mixed oil.

For conducting an efficient flushing operation, the knowledge of mass transfer aspects plays a crucial role. It is important to accurately estimate the oil concentration distribution, flow rate, flow regimes, desired physical properties of the new batch (viscosity, density), pipeline parameters (length, diameter), pressure drop, and friction factor (Riazi 2005). In this work, we study these key predictive parameters to arrive at an optimum flushing operation using CFD (Computational Fluid Dynamics) simulations. Our results will give insights to help industries conduct an effective flushing operation. The knowledge of the key parameters will help to optimize the operation. Furthermore, our application tool will enable the lube oil plant operators to estimate the specific volumes of flush required to meet the desired specifications of a new batch. Thus, our work will improve the environmental, economic, and resource management footprint of the existing processing operations in the lube-oil industries.

References

Dimas, Diovanina, Valéria V. Murata, Sérgio M. S. Neiro, Susana Relvas, and Ana Paula Barbosa-Póvoa. 2018. "Multiproduct Pipeline Scheduling Integrating for Inbound and Outbound Inventory Management." *Computers andamp; Chemical Engineering* 115 (July): 377–96. https://doi.org/10.1016/j.compchemeng.2018.04.025.

Kirschstein, Thomas. 2018. "Planning of Multi-Product Pipelines by Economic Lot Scheduling Models." *European Journal of Operational Research* 264 (1): 327–39. https://doi.org/10.1016/j.ejor.2017.06.014.

Liao, Qi, Pedro M. Castro, Yongtu Liang, and Haoran Zhang. 2019. "Computationally Efficient MILP Model for Scheduling a Branched Multiproduct Pipeline System." *Industrial andamp; Engineering Chemistry Research* 58 (13): 5236–51. https://doi.org/10.1021/acs.iecr.8b06490.

Patrachari, Anirudh R., and Arland H. Johannes. 2012. "A Conceptual Framework to Model Interfacial Contamination in Multiproduct Petroleum Pipelines." *International Journal of Heat and Mass Transfer* 55 (17): 4613–20. https://doi.org/10.1016/j.ijheatmasstransfer.2012.04.017.

Ramanujan, Anirudh. 2008. "Deterministic Models to Explain the Phenomenon of Interfacial Mixing in Refined Products Pipelines," 177.

Riazi, M. R. 2005. *Characterization and Properties of Petroleum Fractions*. W. Conshohocken, PA: ASTM International.

122d Preventing H_2s Poisoning of Pd Membranes for H_2 Purification Using Applied Electric Fields

Daniel J. Rivera, Christopher Muhich

Petroleum refining annually consumes approximately 26 million tons of Hydrogen at a cost of $30 billion. Most of this hydrogen is produced by methane or naphtha reformation, which simultaneously produces several contaminants (H_2S, H_2O, etc.). This H_2 then requires purification which constitutes roughly 1/5 of the total dollar cost and accounts for ~15% of total efficiency losses in the H_2 production process. Theoretically, the best separation process would consist of a single step, which could be achieved by a dense Pd metal membrane. However, the prevalence of contaminants, particularly H_2S, which poisons and corrodes Pd membranes, prevents their widespread adoption in petroleum refining despite their high H_2 flux and selectivity. Here, we use density functional theory to demonstrate a proof of concept for using an applied electric field (AEF) to prevent the decomposition of H_2S on the Pd(111) surface. We find that strong (2 V/Å) AEFs oriented toward and away from the Pd surface similarly reduce the adsorption energy of H_2S but have conflicting effects on the barrier of S-H scission such that a strong field oriented toward the surface reduces the rate of H_2S decomposition. These findings are integrated into a microkinetic analysis which shows no significant surface coverage of Sulfur adatoms at this field strength and orientation. Here, the overall effects of the AEF strength and orientation on the surface electronic density, reaction thermodynamics and kinetics, and surface coverage over time are reported.

122e Understanding Catalyst Failure Mechanisms in Plant Operation

Jin Yang, Vera Santos, Ewa Tocha, David G. Barton, Eric Stangland, Daniel Grohol, van Bloois Stefan, Joost Depicker, Ailene G. Phillips, Michael B Dr Clark, Leadley Stuart, Patrick Heider

Unexpected catalyst failures have strong economic consequences on plant operation. The catalyst life time is determined by various factors, such as loss of catalytic activity and/or selectivity or reaching unsafe operational limits. Solid catalyst deactivation can be grouped into six intrinsic mechanisms: (1) poisoning, (2) fouling, (3) thermal degradation, (4) vapor compound formation and/or leaching accompanied by transport from the catalyst surface or particle, (5) vapor–solid and/or solid–solid reactions, and (6) attrition/crushing. Catalyst degradation under industrial conditions is often due to a combination of mechanical and chemical sources as well as process operation parameters. Thus proper root cause analysis is very complex and often multidimensional fundamental scientific approaches are needed.

This study will show application of microscopy, spectroscopy and development of accelerating aging methodologies on different length scales to understand catalyst failure modes. Herein, challenges of plant catalyst examination will be discussed using as an example CuZnO catalyst. The spent catalyst may not be directly compared with the fresh catalyst because of structure modifications due to reduction and oxidation, thermal stress and mechanical unloading. However, careful interpretation of the results from microscopy analysis enables identification characteristic fingerprints of the failure mechanisms. Origins of different degradation modes for two CuZnO catalysts with very similar composition will be explained (Figure). This work will demonstrate the great potential of post-mortem catalyst structure analysis in understanding the root cause of failure.

122f Strategies for Systematically Optimizing the Operational Management of Lube-Oil Manufacturing and Packaging Facilities

Swapana Jerpoth, Robert Hesketh, C. Stewart Slater, Mariano J. Savelski, Kirti Yenkie

1. Introduction

In lube-oil industries, multiple lube-oil products are consecutively transported in batches through a single pipeline network. During a changeover operation from one product type to another, the lines must be cleaned to ensure the integrity of the new product batch. Lube-oil industries pose many restrictions and the use of any foreign aqueous/non-aqueous based solvents for cleaning/flushing the pipelines is considered a source of contamination to the highly sensitive finished lube-oil products. Hence, the industries use a finished product from the current batch to flush the pipelines from the residues of a previous batch. This leads to the generation of commingled (mixed) oil, which does not match the desired specifications of either of the two batches. The existing flushing operations are controlled by a flush timer. A certain flush time is chosen by an operator based on his previous experience with a particular product. At the end of the chosen flush time, the samples are collected and tested for various physical properties to ensure product quality. If the specifications do not meet the desired range, an additional flushing is conducted. The current procedure mostly relies on trial and error and results in long hold time/ downtime for the lab testing. Moreover, in most scenarios it results in excessive flushing, generating large volumes of commingled oil and leads to high economic losses to these industries. To this end, our work aims to address the existing drawbacks and explore alternative operational procedures aiming to minimize commingled oil volumes and improve the economic and environmental footprint of the flushing operations in lube-oil industries.

In recent years, optimization techniques have gained growing interest in the petroleum industry. MINLP models have been widely used for developing systematic scheduling plans and calculating commingled oil volumes for multiproduct pipelines (Chen et al. 2021; Abdellaoui et al. 2018). Major oil companies are developing models based on empirical correlations that are only applicable to their respective pipeline networks. There is no widely accepted correlation that can be used in all actual scenarios (He et al. 2018). Hence, based on these fundamentals and existing literature, we developed an idea of employing optimization techniques and formulating the flushing operation as an optimal control problem.

2. Mathematical Models

In this work, we investigate the lube-oil processing pipelines of our partnered industry, one of the world leaders in manufacturing lubricants. At this facility, the effectiveness of a flush is tested by collecting samples at the end of a designated flush time and analyzing them for their composition through various laboratory tests. These include viscosity, density, color, water content, and spectrophotometry. Viscosity is considered the preliminary and most crucial differentiable property for ensuring the product quality of individual batches. However, the conventional methods of testing viscosity require a long-running time leading to extended operational downtime. Thus, this work addresses the existing drawback and enables better in-line controllability of the flushing operation by developing models for predicting the viscosity of the lube-oil blends in real-time. The viscosity blending correlations recommended by API-TDB were combined with the component balance equations for the pipeline networks at our partnered lube-oil facility (Riazi 2005; Roegiers and Zhmud 2011). Our developed models were validated against experimental data that was collected through systematically designed and performed plant experiments. The agreement within ±5% error between the experimental and simulated data confirmed that our models hold value.

3. Solution Strategy

The flushing involves controlling a dynamic system, i.e., the system that evolves. Hence this work uses optimal control principles for optimizing the existing operations (Diwekar 2008). The developed problem was solved using two solution techniques viz. maximum principle and non-linear programming (Yenkie and Diwekar 2014b; 2014a). Our solution algorithm was developed using the programming language of MATLAB and the method involved the analysis of time-dependent first-order and second-order differential equations for generating an optimum flushing time. The results of the two solution methods were compared and the pros and cons are discussed.

4. Conclusion and Future Work

Our work highlights the advantages of using optimization techniques and API recommended property prediction correlations, for eliminating the drawbacks of the existing flushing operations in the lube-oil industries. We present methods that can enable industrial stakeholders to make informed decisions and thereby reduce the operational costs and downtime of these processes. This work will thus serve as an excellent starting point in improving the environmental and economic footprints of these operations.

5. References

Abdellaoui, Wassila, Asma Berrichi, Djamel Bennacer, Fouad Maliki, and Latéfa Ghomri. 2018. "Optimal Scheduling of Multiproduct Pipeline System Using MILP Continuous Approach." In *Computational Intelligence and Its Applications*, edited by Abdelmalek Amine, Malek Mouhoub, Otmane Ait Mohamed, and Bachir Djebbar, 411–20. IFIP Advances in Information and Communication Technology. Cham: Springer International Publishing. https://doi.org/10.1007/978-3-319-89743-1_36.

Chen, Lei, Ziyun Yuan, JianXin Xu, Jingyang Gao, Yuhan Zhang, and Gang Liu. 2021. "A Novel Predictive Model of Mixed Oil Length of Products Pipeline Driven by Traditional Model and Data." *Journal of Petroleum Science and Engineering* 205 (October): 108787. https://doi.org/10.1016/j.petrol.2021.108787.

Diwekar, Urmila. 2008. *Introduction to Applied Optimization*. Springer.

He, Guoxi, Mohan Lin, Baoying Wang, Yongtu Liang, and Qiyu Huang. 2018. "Experimental and Numerical Research on the Axial and Radial Concentration Distribution Feature of Miscible Fluid Interfacial Mixing Process in Products Pipeline for Industrial Applications." *International Journal of Heat and Mass Transfer* 127 (December): 728–45. https://doi.org/10.1016/j.ijheatmasstransfer.2018.08.080.

Riazi, M. R. 2005. *Characterization and Properties of Petroleum Fractions*. W. Conshohocken, PA: ASTM International.

Roegiers, Michel, and Boris Zhmud. 2011. "Property Blending Relationships for Binary Mixtures of Mineral Oil and Elektrionised Vegetable Oil: Viscosity, Solvent Power, and Seal Compatibility Index." *Lubrication Science* 23 (6): 263–78. https://doi.org/10.1002/ls.154.

Yenkie, Kirti M., and Urmila Diwekar. 2014a. "Optimal Control for Predicting Customized Drug Dosage for Superovulation Stage of in Vitro Fertilization." *Journal of Theoretical Biology* 355

(August): 219–28. https://doi.org/10.1016/j.jtbi.2014.04.013.

Yenkie, Kirti M., and Urmila M. Diwekar. 2014b. “Comparison of Different Methods for Predicting Customized Drug Dosage in Superovulation Stage of In-Vitro Fertilization.” *Computers andamp; Chemical Engineering* 71 (December): 708–14. https://doi.org/10.1016/j.compchemeng.2014.07.021.

122g Demulsification of Tight Crude Oil-Water Emulsions Under Microwave Radiation in Presence of Nanohybrid Hydrophilic Structures

Prasad Pawar, Pankti Joshi, Clayton Jeffryes

The main objective of this study is to use microwaves to develop an energy efficient and economical separation method to demulsify stable crude oil and water emulsions. This separation method will be further enhanced by using reusable, *in situ* materials to further destabilize the oil-water interface in the emulsions. The crude oil emulsions are formed by high shear mixing of process water and crude oil during pumping, desalting and transportation through wellbores and pipelines and are stabilized by emulsifying agents that form interfacial films at the oil-water interface. These surface-active species, such as asphaltenes, resins, solids, waxes, or commercially available chemical emulsifiers present in the crude, are responsible for the formation of stable emulsions. Tight crude oil emulsions, characterized by suspended droplets on the order of submicron to tens of microns, often cause serious problems to the wellbores and pipelines such as clogging, corrosion, and pump failures. The most common method widely used is the chemical demulsification method by addition of chemicals called demulsifiers. These chemicals neutralize the stabilizing effects of the emulsifiers that stabilize the emulsions. The surface active demulsifiers move to the oil/water interface and weaken or break the rigid films formed by the asphaltenes, resins, and waxes present in the emulsion. This enhances the coalescence of the dispersed water droplets. With increasing demand for nanomaterials in various fields, they have also gained interest in the petroleum industry. Graphene oxide (GO) has oxygen containing functional groups like epoxide, hydroxyl, carbonyl, and carboxyl groups. The presence of these oxygen-containing functional groups provides potential advantages for using GO in numerous applications. The polar oxygen functional groups in GO sheet make it strongly hydrophilic, which give GO good dispersibility in many solvents, particularly in water. It is hydrophilic and forms stable suspension in water and other solvents. Polyethylene glycol is a hydrophilic chemical demulsifier used for crude oil water separation.

Here we have prepared a hydrophilic nanohybrid structure (GO-PEG) using graphene oxide and polyethylene glycol (PEG). Fourier transform infrared (FTIR) spectroscopy and X-ray diffraction were used to characterize the prepared GO-PEG structure. To simulate field conditions, we mixed two crude oils to prepare samples with American Petroleum Institute (API) gravities between 28.9° and 46.0°. These crudes were mixed with water to create emulsions containing between 20% and 80% water. This array of Water-in-Oil (W/O) and Oil-in-Water (O/W) emulsions with different densities and viscosities were mixed at varying shear rates. To study their individual effects on demulsification, GO, PEG, and GO-PEG were added to these emulsions at different concentrations of 0, 200, 500, and 1000 mg/L. A series of batch demulsification runs of these crude oil emulsions with GO, PEG, GO-PEG additives were done by exposure to microwaves under a well-defined set of parameters such as temperature, power, time, and salinity. To determine the efficiency of phase separation after treatment, final oil-phase water content was measured using Karl Fischer Titration. Change in droplet size of dispersed phase was analyzed using the optical microscopy images of the emulsions before and after treatment. Energy input analysis was performed from the Power-Time data obtained from each demulsification run. A separation efficiency model was developed based on the energy and mass balance study for the microwave demulsification method.

SCREENING AND INVESTIGATION ON INHIBITION OF SEDIMENT FORMATION IN A KUWAIT LIGHT CRUDE OIL BY COMMERCIAL ADDITIVES WITH SOME GUIDELINES FOR FIELD APPLICATIONS

Ali Qubian[1], Abdulaziz S. Abbas[1], Noura Al-Khedhair[1], Jose Peres[1], O. Alomair[2], Dicho Stratiev[3,4], Ivelina K. Shishkova[4] and M R Riaz[2,5], (1) Kuwait Oil Company, Ahmadi, Kuwait, (2) Kuwait University, Safat, Kuwait, (3) Bulgarian Academy of Science, Sofia, Bulgaria, (4)Lukoil-Neftohim, Burgas, Bulgaria, (5) Oil and Gas Inc., Montreal, QC, Canada

Abstract

Crude oil is a very complex mixture, consisting of myriad components, that during changes in operating conditions such as pressure, temperature and composition of the oil can trigger deposit formation during oil production operations. The deposit formation is a complex phenomenon that is mainly provoked by asphaltene aggregation and precipitation and/or by precipitation of wax or paraffin. The precipitation of asphaltene, and waxes occurs when crude oil characteristics change as a consequence of pressure, temperature variations, and/or chemical modifications, etc. Asphaltene characterization, as a pre-screening for appropriate inhibitors, has been conducted using asphaltene phase diagrams (APD). For this purpose, two software were developed in both Matlab and Excel codes to predict stable and unstable regions for this crude oil sample. The APD generated from these softwares favorably are compared with phase behavior of other oil samples available in the literature. The aim of this study was to verify the feasibility of suppressing the process of sediment formation in a light crude oil by treating this crude oil with commercial antifouling additives. Basic properties such as density and distillation (ASTM 7169), sulfur content, pour point, sediment, and SARA analysis were determined. The solvent power and the critical solvent power of the crude were determined as well. Seventeen chemicals from five global companies were screened for testing to inhibit the precipitation. Optimum concentration and the amount of reduction in precipitation were determined for all of these chemicals to identify the most suitable chemicals. Some recommendations are given for the field application of such chemicals.

Introduction

Asphaltene precipitation can negatively affect the oil recovery and refining processes from its early stage in the reservoir, during Enhanced Oil Recovery (EOR), to flow of produced oil in the production well, and surface facilities. Through adsorption of crude oil polar components onto surfaces, the asphaltene can alter wettability. It will also block pore spaces, resulting in reduced local permeability and therefore, reduced oil production rates. The asphaltene deposition can also occur in the production well where the pressure drop is maximum, and the thickness of deposited asphaltene increases over time and this may cause decrease in production rate. Also it is reported that increase in asphaltene precipitation has been observed in sections of the well with increase in turbulence in flow. Asphaltenes are complex molecules with molecular weights ranging from 1000 to 5000 g/mol and density of about 1100-1250 kgm^{-3}. Asphaltene molecules contain some heteroatoms such as nitrogen, sulfur and oxygen. They are mainly aromatics and may precipitate at certain thermodynamic conditions such as temperature, pressure and oil composition. Asphaltene colloid formation, flocculation and precipitation processes have been studied and reported in the literature (1-4).

Presence of resins plays an important role in asphaltene stability or its precipitation. During gas injection into a reservoir for enhanced oil recovery processes, the composition of oil changes and consequently precipitation may occur. In addition to the composition of crude oil, type and amount of injected gas, temperature and pressure, flow characteristics and properties of the production well will affect asphaltene precipitation. Because of the complex nature of asphaltene the phenomenon of its precipitation never fully understood by the researchers despite extensive research conducted in this area over the last several decades. Asphaltene stability test methods and screening of chemicals for asphaltene inhibition in laboratory have been reported by some researchers (5-8). The three main questions for the industry regarding asphaltene deposition are: i-when it happens, ii-how much precipitation occurs and iii-how to prevent or to reduce the amount of precipitation.

In this paper the thermodynamic model based on PC-SAFT equation of state previously applied to Kuwait oils is used to examine the phase behavior and instability of a light crude oil from Marrat reservoir in Kuwait (9). In addition a group of chemicals manufactured by five global companies used as commercial asphaltene inhibitor were investigated for their performance on this crude oil sample and the most suitable chemicals were identified and recommended for possible field application.

Laboratory Data and Stability of Oil Sample

Bottom Hole Oil Sample and PVT Data

The oil sample was taken by the operating company from onshore well MN-27 in Marrat formation located in the Minagish field in West of Kuwait also known as WK oil sample. This was the main sample used for the experiments, testing and evaluation. Three samples of each 500 ml were taken on June 5, 2018 at a depth of 14,000 ft. The bottom hole pressure and temperature were 4063 psi and 242 °F, respectively. This oil sample was used in the petroleum research facility laboratory at Kuwait University to measure basic PVT data as given in Tables 1 and 2. The bubble point was determined from constant mass experiment (CME) conducted at 242 °F. The composition of oil sample was determined from PVT and subsequent GC analysis and is given in Table 3 (for the C_{7+} fraction: MW=211.97 and SG=0.842).

Crude Analysis and Stability Test

In addition to the live oil sample, 10 L of dead crude oil from well MN-27 was collected by the gathering center of KOC. Basic measured properties are given in this section. Density at 20 °C: 0.8277 g /cm^3, C7 asphaltenes: 1.6 wt.%, sulfur content: 1.049 wt.% and API gravity: 39.3. Boiling point and API gravity distribution of the crude are shown in Figure 1.

One key question regarding asphaltene issue is to know its stability under the operating conditions. If the oil is unstable, asphaltene can precipitate and if stable it is soluble in oil. This can be determined by using Turbiscan instrument with a simple experiment. However, there are some simple analytical methods that can be used to determine oil stability. The line between stable and unstable oil is not definite and there could be a region of uncertainty known as mildly stable or mildly unstable. One simple method is to determine a parameter known as Colloidal Instability Index (CII) defined based on SARA analysis as the ratio of (Saturates%+ Asphaltenes%)/(Aromatics%+Resins%). If CII is less than 0.7 the oil is stable and if greater than 0.9 it is unstable. If CII is between 0.7 and 0.9 the oil is mildly unstable (7). The asphaltene /resin ratio, on the other hand, relates to the measure of colloidal stability of the asphaltenes. If this ratio is greater than 0.3 the oil is unstable (8). Oils with higher resin content are more stable with addition of a solvent such as n-C5 or n-C7. For the crude MNL-27 the CII is 2.03 which is greater than 0.9 and hence the oil is unstable. This is also shown graphically in Figure 2.

Table 1. Measured SARA Analysis, Bubble Point and Density for Oil Sample MN-27

Sara Fraction	Weight %		Reservoir Initial Pressure, psig	9300
Saturates	65.46		Bubble Point Pressure, psia	2271.6
Aromatics	28.27		Density at Reservoir Conditions, kg/m^3	640
Resins	4.69			
Asphaltenes	1.58		Reservoir Temperature, °F	242

Table 2. General Properties at Standard Conditions for Oil Sample MN-27 (West Kuwait)

Sample Volume Used at Res. T&P, mL	67.475		UAOP, psia at 242 °F UAOP, psia at 185 °F	5200 5050
Flashed Liquid Volume at STP, mL	47.07		Density at 60 F, kg/m^3	823.8
GOR, scf/bbl	675.46		Molecular Wt.	194
Basic Sediment & Water Content, wt.%	0.0		AP Gravity	40.11
Asphaltenes	1.58		Viscosity at 60 °F, cP	4.9067

Table 3. Composition of Oil Sample MN-27 (West Kuwait)

Component	CO_2	N_2	H_2S	C_1	C_2	C_3	nC_4	iC_4	nC_5	iC_5	C_6	C_{7+}
Mol%	2.03	0.12	1.91	27.47	12.68	8.23	3.1	0.9	3.41	2.95	5.36	31.84

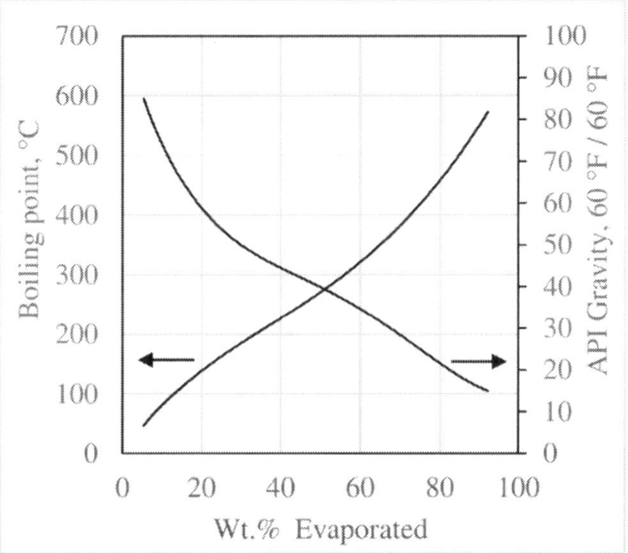

Figure 1: API gravity and boiling point distribution of crude oil sample

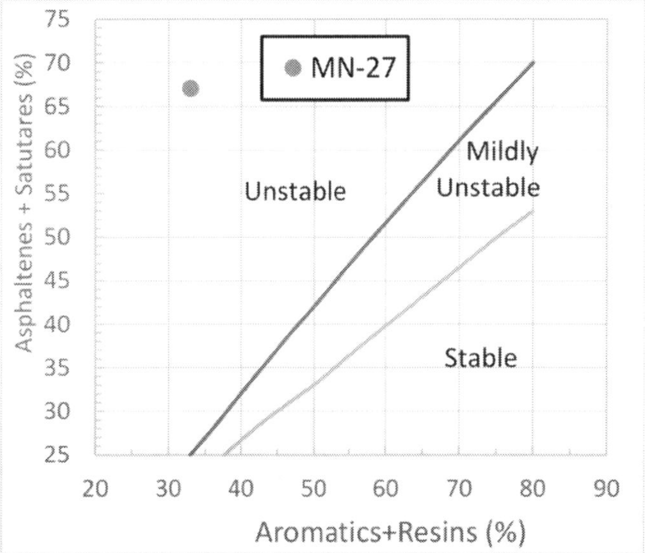

Figure 2. Stability Test for Crude Sample MN-27

Region of Asphaltene Instability Using an Equation of State

It is important to determine the region in the phase diagram that asphaltene precipitation can occur. Such information is vital to check under certain operational conditions of temperature, pressure and composition. This can be done through asphaltene phase diagram (APD). Generation of APD is the key to determine under what conditions asphaltene formation occurs. A thermodynamic model based on PC-SAFT equation of state has been successfully applied to asphaltenic oils for phase behavior calculations (10). Its modeling and application to crude oils from Kuwaiti fields was presented previously (9). Programs with both Matlab and Excel codes were developed and tested literature data. With use of data in Tables 1-3, the EOS parameters were optimized and the phase envelope for the asphaltene instability region was determined as shown in Figure 3. If the operating condition is above or below the Asphaltene Onset Pressures (AOP) the oil is stable.

Screening of Chemical Inhibitors for Asphaltene Inhibition

The problem with asphaltene precipitation in flow assurance can be reduced in three different ways: 1) to prevent it from happening, 2) to reduce the extent of precipitation, and 3) to dissolve deposited asphaltene. The extent of asphaltene precipitation can be controlled by not allowing the size of asphaltene aggregate to grow. For example, the adsorption of non-ionic dispersants onto the surface of asphaltene particle can avoid its growth and therefore, will limit the size to aggregate, which in turn will allow the asphaltene to be carried along the oil phase.

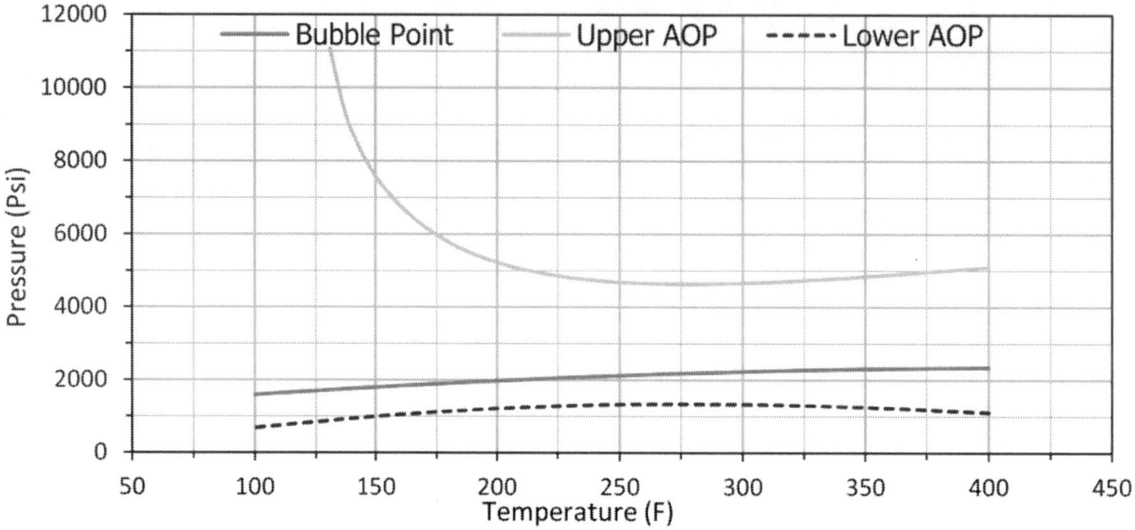

Figure 3. Asphaltene phase diagram for oil sample M-27

There are clearly two groups of chemical additives that can prevent asphaltene deposition known as asphaltene dispersants and inhibitors. Examples of non-polymeric dispersants are: very low polarity alkylaromatics or alkylaryl sulfonic acids. Examples of asphaltene inhibitors are: alkylphenol/aldehyde resins and similar sulphonated resins, polyolefin esters, amides, or imides with alkyl, alkylenephenyl, or alkylenepyridyl functional groups and alkenyl/vinyl pyrrolidone copolymers. These chemicals are mainly used in wellbores and surface equipment and increase the asphaltene stability under wider operation conditions (11). These chemicals are oil-specific, for example presence of nitrogen in asphaltene can interact with the polymeric inhibitors containing H^+ atoms such as hydroxyl groups. In general asphaltene inhibitors are polymeric type chemicals and to be effective we must reach a certain critical concentration while dispersants act almost proportionately to concentration. Use of live oils and dead oils in testing such chemicals may give different results, however, tests with dead oil can be used for general screening of chemicals since the trends of effectiveness is similar to those with live oils (7).

Commercial Chemical Inhibitors Used In this Study

A total of 24 inhibitors available in our research laboratory were used for testing. Types, base and nature of some of these inhibitors are given in Table 4. These chemicals were purchased from 5 major companies in the world that manufacture chemicals and additives for inhibition, dispersion and dissolving (dissolution) of asphaltene and other types of solid deposition from petroleum fluids for applications in the industry. Some of these chemicals when initially tested with our crude did not show a positive performance, in fact some of them promoted further deposition rather than inhibition of precipitates. Commercial companies do not reveal the exact composition of their chemicals and only introduce by their own chemical codes. These code numbers are frequently changed with time and from one client to another that use the chemicals. For this reason, we do not know the exact composition of chemicals tested.

Ultimately 17 chemicals were selected for testing which they all fall in the group types given in Table 4.

Table 4. Type and base of chemicals used in our chemical screening process

<table>
<tr><td>Amines in aromatic solvents</td><td>Alkene C_{24} – C_{36}, alpha-polymers with maleic anhydride</td></tr>
<tr><td>Polyisobutylene Succinimide</td><td rowspan="2">Phosphoric acid,
2-ethylhexyl ester | $C_8H_{17}O_4P$-2</td></tr>
<tr><td>Polymer in aromatic solvents</td></tr>
<tr><td>Aromatic Hydrocarbons</td><td>Organic acid derivative</td></tr>
<tr><td>Oxidized light distillates</td><td>1,2,4 Trimethylbenzene</td></tr>
<tr><td>Phosphoric acid, 2-ethyl-hexyl ester</td><td>PAH hydrocarbons in solvents</td></tr>
</table>

Chemical Screening Test Results

Asphaltene Dispersant Test (ADT) method was used to determine performance of each chemical (12,13). For each test 1.5 gram of oil was diluted by 20 gr of n-heptane before a chemical additive was added. Results are given in terms of additive concentration versus amount of the deposit as well as percent reduction in the deposit from its initial amount before addition of additive as shown in Figures 4 and 5 for one of best performing chemicals for this particular oil sample.

In Figure 5, negative % means reduction in deposit and +% (above 0 line) means % increase in deposit and the chemical is promoting deposition rate. For this inhibitor at 500 ppm and higher concentrations it can reduce the deposition by 75%. The optimum chemical concentration for this oil sample was 500 ppm. Some chemicals promoted deposition at low concentration while at higher concentration deposition was reduced as shown in Figure 6. On this basis the best performing chemical can be selected.

Field Applications

It is believed that the amount of asphaltene deposition in wellbore is less than those given in the experiments conducted in laboratory. In the wellbore temperature is higher than room temperature and there is flow of oil in contrast with lab conditions. Both of these factors can cause a reduction in asphaltene precipitation under other similar condition with laboratory condition. Accordingly, it is expected that amount of inhibitor used in the well bore will be less than those measured in laboratory at room temperature. Successful asphaltene management requires consideration across the entire production cycle from the reservoir to flowline and process infrastructure, and ultimately to oil export. In general, application of asphaltene inhibitors is best as far upstream as reasonably possible with the preferred injection point being downhole. Asphaltene destabilization is mainly pressure driven, thus the injection of an asphaltene inhibitor should happen at the highest possible pressure before the destabilization occurs. Application of the asphaltene inhibitor should be continuous at a steady dose rate to be optimized in the field. Fluid composition, flow rate, pressure drop and asphaltene stability should be frequently monitored for best application of a chemical injection for asphaltene deposition control.

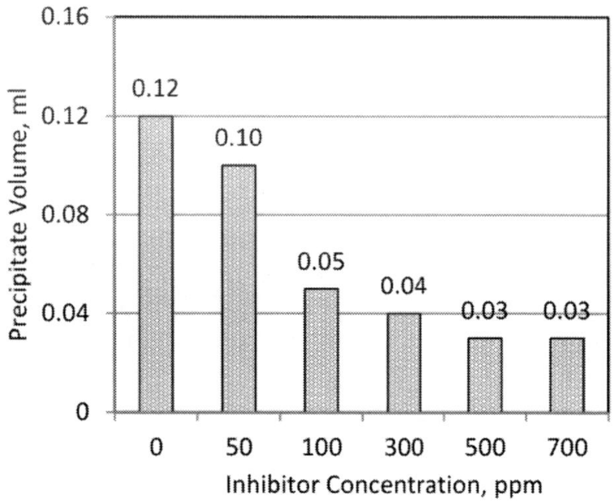

Figure 4. Amount of deposit versus treating rate of inhibitor

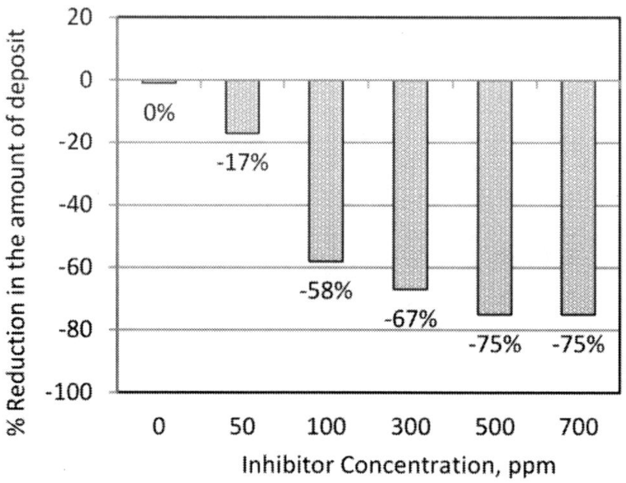

Figure 5. % Reduction in deposit versus treating rate of a chemical recommended

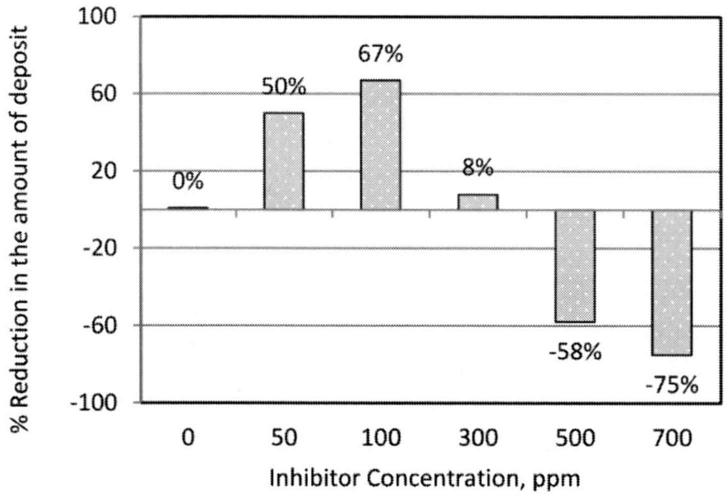

Figure 6. % Reduction in deposit versus treating rate of a chemical not recommended

References

1. Akbarzadeh, K., A. Dhillon, W.Y. Svrcek, H.W. Yarranton (2004), "Methodology for the Characterization and Modeling of Asphaltene Precipitation from Heavy Oils Diluted with n-Alkanes", Energy and Fuels, 18(5), pp. 1434-1441.
2. Mohammed, Isah, M. Mahmoud, D. Al Shehri, A. El-Husseiny and O. Alade (2021), "Asphaltene Precipitation and Deposition: A Critical Review, J. Pet. Sci. Eng. 197, 107956.
3. Riazi, M.R. (2005), "Characterization and Properties of Petroleum Fractions", ASTM International: MNL50, Conshohocken, PA.
4. Mousavi-Dehghani, S. A., M. R. Riazi, M. Vafaie-Sefti and G. A. Mansoori (2004), "An Analysis of Methods for Determination of Onsets of Asphaltene Phase Separations," J. Pet. Sci. Eng., 42, pp. 145-156.
5. Pereira, V.J., L.L.O. Setaro, G.M.N. Costa and S.A.B. Viera de Melo (2017), "Evaluation and Improvement of Screening Methods Applied to Asphaltene Precipitation," Energy and Fuels, 31(4), pp. 3380 -3391.
6. De Boer, R.B., K. Leelooyer, M.R.P. Eigner, and A.R.D. van Bergen (1995), "Screening of Crude Oils for Asphalt Precipitation: Theory, Practice, and the Selection of Inhibitors," SPE Prod & Fac. 10 (01), pp. 55–61.
7. Yen, A., Y.R. Yin and S. Asomaning (2001), "Evaluating Asphaltene Inhibitors: Laboratory Tests and Field Studies", SPE International Symposium on Oilfield Chemistry, Houston, Texas, February.
8. Jamaluddin, A.K.M., J. Creek, C.S.Kabir, J.D. McFadden, D. Dcruz, J. Manakalathil, N. Joshi and B. Ross (2002), "Laboratory techniques to measure thermodynamic asphaltene instability," J. Can. Pet. Technol., 41 (7), pp. 44-52.
9. Riazi, M.R. and O. Alomair (2021), "Modeling and Prediction of Asphaltene Precipitation for Two Crude Oil Samples from Kuwaiti Fields," Annual AIChE Meeting, Virtual, November.
10. Kurup, A.S., J. Wang, H. J. Subramani, J. Buckley, J. L. Creek, and W. G. Chapman (2012), "Revisiting asphaltene deposition tool (ADEPT): field application," *Energy and Fuels,* 26(9), pp. 5702-5710.
11. Kelland, M. A. (2009), "Production Chemicals for the Oil and Gas Industry", Taylor & Francis, CRC Press, NY.
12. Da Silva Ramos, A.C., Lilian Haraguchi, Fabio R. Notrispe, Watson Loh and Rahman S. Mohamed (2001), "Interfacial and Colloidal Behavior of Asphaltenes Obtained from Brazilian Crude Oils," J. Pet. Sci. Eng. 32, pp. 201 – 216.
13. Stratiev, D., R. Dinkov, I. Shishkova, I. Sharafutdinov, Natalia Ivanova, M. Mitkova, D. Yordanov, N. Rudnev, K. Stanulov, A. Artemiev, I. Barova and B. Chushkov (2016), "What Is Behind the High Values of Hot Filtration Test of the Ebullated Bed Residue H-Oil Hydrocracker Residual Oils?" *Energy & Fuels*, 30, pp. 7037–7054.

122i Crude Assays - the Distillation

Andrew Sloley

Crude Assays:

The Distillation

Andrew W. Sloley

DGI

Bellingham, WA, USA

Since the first oil well was drilled in Titusville, PA in 1859 the refining industry has continuously worked to cope with the fundamental question of 'What is that crude worth?' The value of a specific crude varies with its processing costs and the values of its possible products. Over time, the crude assay has been developed to answer this question. The elements of a crude assay include a mix of vaporization characteristics and properties of possible products that might be made from the crude. The key points for the first processing unit in the oil refinery, the Crude Distillation Unit, revolve around the distillation curve and properties related to corrosion and fouling.

This paper reviews the development of the distillation curve. Topics include the current status of distillation curve methods and their interpretation and use. Near-future developments, and their possible use are also discussed. Discussion include how distillation curve use interacts with prediction of molecular weights from density or other methods.

122j Mechanistic Investigation of Asphaltene-Induced Fouling of Heat Exchangers

Yousef Al-Shamlan

Asphaltene precipitation and its subsequent deposition is a growing problem for the oil and gas industry. Not only does it hinder the output of oilfields, but also causes significant issues in downstream refining. The blending of incompatible oils and operations at high surface temperatures are some of the key causes of the instability of asphaltenes, which may lead to fouling and reduction in performance for refinery equipment, such as heat exchangers. In this work, the combination of state-of-the-art laboratory techniques, and advanced modeling methods are used to better understand and predict the occurrence and the magnitude of asphaltene-induced fouling of heat exchangers.

The PC-SAFT model which has proven its success in upstream applications has been adapted to further explore asphaltene phase behavior and deposition tendencies at refinery conditions. A novel pilot-scale setup has also been developed to investigate the tendencies of heat exchangers to foul at high temperatures by quantifying the amount of deposited material at the end of each experiment. The modeling and lab results obtained from this study will be compared to commercial fouling rigs and field conditions. With this study, we aim to contribute to the development of best practices to mitigate asphaltene-induced fouling that is commonly found in refineries around the world.

122k Dual Functional Catalysts for Cracking of Light Cycle Oil into Monoaromatics Using a Hydrogen Donor

Akshata Ramteke, Dwijraj Mhatre, Divesh Bhatia, Kamal Pant

Globally, declining growth for fuels and rapid growth for petrochemicals has led oil refiners to look alternatives for converting crude to chemicals. In such a scenario, transformation of low value and low demand fuels to high value and high demand petrochemicals is highly benign. Light cycle oil (LCO) is obtained as a middle distillate in the diesel cut of the fluidized catalytic cracking (FCC) unit of petroleum refineries. Di and tri aromatics are the predominant molecular species in light cycle oil (LCO), that need to be converted to monoaromatics. However, conversion of LCO under FCC conditions is very low and remains least explored in the literature. The focus of this work is the conversion of diaromatics present in LCO to monoaromatics such as benzene, toluene, and xylene (BTX) under FCC conditions. Catalytic cracking of LCO and its model compounds was performed over monometallic (Co, Mo) and bimetallic (CoMo) catalysts supported over Beta and Y zeolite. Instead of hydrogen, n-hexadecane was used as the hydrogen donor under FCC conditions, which has the potential to reduce the process costs. The catalysts were synthesized by co-impregnation, and their performance was evaluated in a fixed-bed micro activity testing (MAT) reactor setup at 500 ^{0}C and 1 atm. CoMo/Beta catalyst resulted in the maximum monoaromatics yield of 32.7% at 73.8% conversion of LCO in the presence of n-hexadecane, whereas in the absence of n-hexadecane, the corresponding monoaromatics yield was 25.3% at 48.6% LCO conversion. Thus, n-hexadecane serves as an effective hydrogen donor to promote the hydrogen transfer reactions and results in a higher yield of the desired product. The amount of coke also decreased in the presence of n-hexadecane. The monoaromatics yield for various catalysts followed the order: CoMo/Beta > CoMo/Y > Mo/Beta > Co/Beta > Mo/Y > Co/Y > Beta > Y, whereas the coke and diaromatics formation followed the reverse order, with CoMo/Beta resulting in the minimum coke and diaromatics formation . The improved catalytic activity of the bimetallic catalysts as compared to monometallic catalysts and commercial zeolites is attributed to the synergistic effect between Co and Mo. Metallic dispersion and interaction of Co and Mo with the support was confirmed using XRD, Raman, FTIR, UV-DRS, FE-SEM, HR-TEM, and XPS.

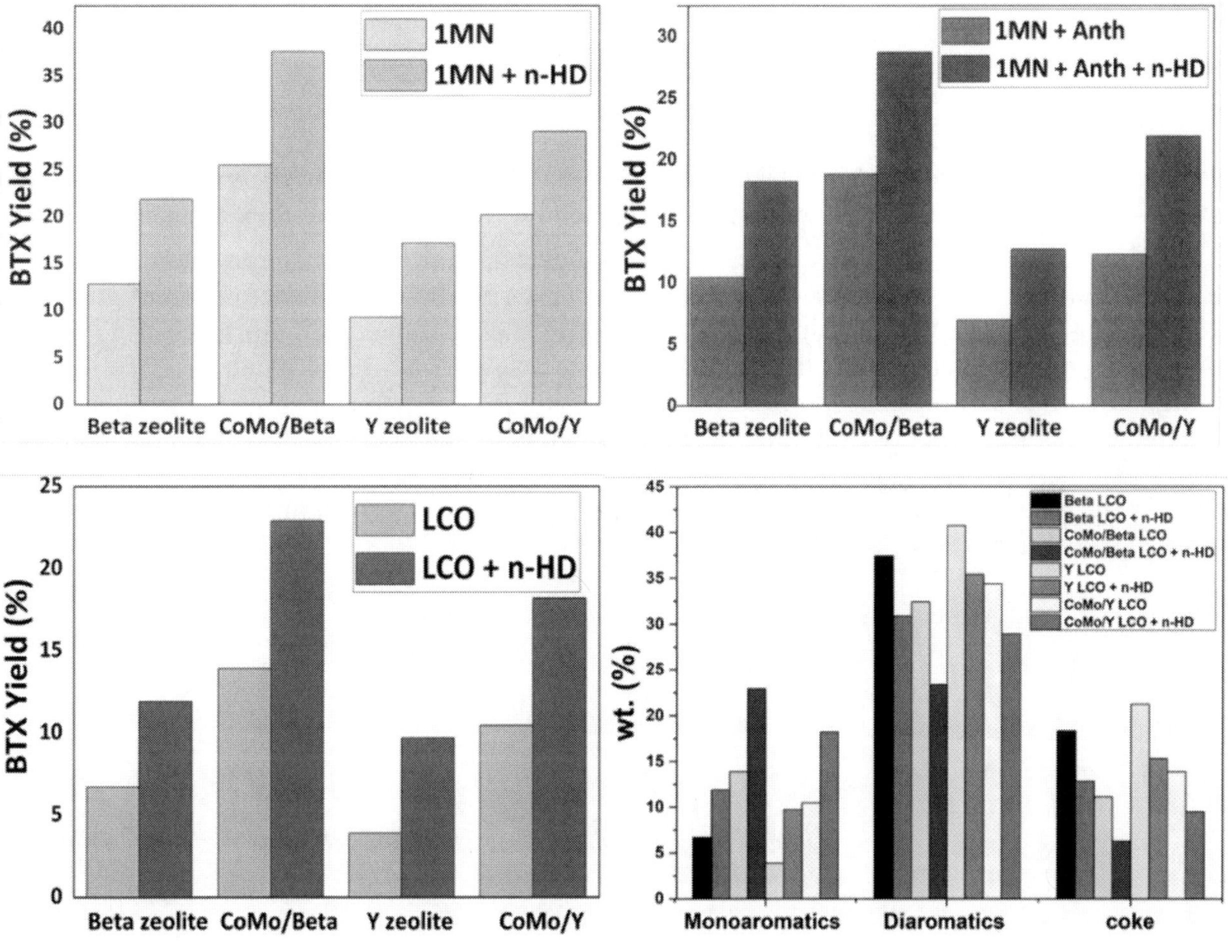

1MN
1MN + n-HD
BTX Yield (%)
Beta zeolite
CoMo/Beta
Y zeolite
CoMo/Y
1MN + Anth
1MN + Anth + n-HD
BTX Yield (%)
Beta zeolite
CoMo/Beta
Y zeolite
CoMo/Y
LCO
LCO + n-HD
BTX Yield (%)
Beta zeolite
CoMo/Beta
Y zeolite
CoMo/Y
Beta LCO
Beta LCO + n-HD
CoMo/Beta LCO
CoMo/Beta LCO + n-HD
Y LCO
Y LCO + n-HD
CoMo/Y LCO
CoMo/Y LCO + n-HD
wt. (%)
Monoaromatics
Diaromatics
coke

123a A Co-Axial Microtubular Electrochemical Flow Cell for High-Power Flow Batteries and Electrolyzers

Nian Liu

Electrochemical flow cells, where electron exchange takes place, is a central component in flow batteries and electrolyzers. Improving the volumetric power density of the flow cell (W/L_{cell}) can reduce the size and cost of flow batteries and electrolyzers. While significant progress has been made on electrode and membrane materials, conventional flow cells based on planar cell configuration exhibit a large cell size with multiple bulky accessories such as flow distributors, resulting in low volumetric power density. In this talk, I will introduce a new co-axial microtubular (CAMT) flow cell configuration that significantly improves volumetric power density by reducing the membrane-to-membrane distance by almost 100 times and eliminating completely the bulky flow distributors. The CAMT cell has been applied to flow batteries and electrolyzers with various redox chemistry. The CAMT flow cell represents a device-level revolution to enhance the volumetric power of flow batteries and electrolyzers.

(also see our publication: DOI: 10.26434/chemrxiv-2022-d8sgv)

123b High Performance Direct Methanol Fuel Cells Incorporating pH-Gradient-Enabled Microscale Bipolar Interfaces

Kritika Sharma, Shrihari Sankarasubramanian, Zhongyang Wang, Vijay Ramani

Herein we report a direct methanol – hydrogen peroxide fuel cell (DMHPFC) operated with a sustained pH-gradient between an alkaline methanol fueled anode and an acidic hydrogen peroxide fed cathode. The DMHPFC produced *ca* 0.5 W cm^{-2} peak power density at 0.85V with an open circuit voltage (OCV) of 1.65V. Direct methanol fuel cells (DMFCs[1,2]) are an important alternative to hydrogen-fed polymer electrode membrane fuel cell (PEMFCs) and anion exchange membrane fuel cells (AEMFCs) due to methanol's high energy density compared to that of hydrogen. A unit volume of Hydrogen gas stored at a pressure of approximately 69 MPa corresponds to volume-specific energy density of roughly 2.1 kWh/l while 39% volume of aqueous methanol and 41% volume of aqueous hydrogen peroxide corresponds to volume-specific energy density of 9.2 kWh/l[3]. So, DMHPFC's energy density is approximately four times higher than hydrogen fuel cell's energy density and is comparable to gasoline which contains roughly 9.2 kWh/l of available bond energy. Further, methanol exhibits good electrochemical activity, is consistently available, is biodegradable and is relatively cheap[4]. However, DMFCs have not been a viable alternative to hydrogen fuel cells due to their low power density caused by the sluggish kinetics of both electrode reactions[5]. In this work, we significantly improve the kinetics of the methanol oxidation reaction (MOR) by moving to alkaline media from acidic media[6]. To address the inferior kinetics of the oxygen reduction reaction (ORR) at the cathode, hydrogen peroxide is used as the oxidant in place of oxygen[7,8]. The theoretical potential for MOR in basic medium is lower than that in an acidic medium (-0.78 V vs. -0.02 V) and the theoretical potential for the reduction of hydrogen peroxide is higher than that for oxygen reduction (1.77 V vs. 1.23 V in acidic medium). This results in the highest theoretically possible cell voltage for cells with alkaline anodes and acidic cathodes.

We achieved sustained operation of the DMHPFC with an alkaline anode and an acidic cathode by incorporating our pH-gradient-enabled microscale bipolar interface (PMBI). We have previously demonstrated that the PMBI can sustain a pH-gradient between the anode and the cathode of a direct borohydride fuel cell (DBFC) and enable sustained high power operation[9–12]. Translating the same technology to DMHPFC, we demonstrate that the judicious selection of the electrode pH and inter-electrode pH-gradients enabled by the PMBI leads to enhanced performance of CH_3OH/H_2O_2 fuel cells compared with CH_3OH/O_2 fuel cells. Similar to methanol, liquid fuels like ethanol, propanol, glycerol also have lower theoretical potential in alkaline medium. Our work paves the way for such other liquid fuel cells with hydrogen peroxide as the oxidant and enabled with PMBI resulting in significantly improved performance.

References:

1. H. Ahmad, S. K. Kamarudin, U. A. Hasran, and W. R. W. Daud, *Int. J. Hydrogen Energy*, **35**, 2160–2175 (2010) http://dx.doi.org/10.1016/j.ijhydene.2009.12.054.
2. C. Zhao, H. Lin, and H. Na, *Int. J. Hydrogen Energy*, **35**, 2176–2182 (2010) http://dx.doi.org/10.1016/j.ijhydene.2009.12.149.
3. D. N. Prater and J. J. Rusek, *Appl. Energy*, **74**, 135–140 (2003).
4. M. Nacef and A. M. Affoune, *Int. J. Hydrogen Energy*, **36**, 4208–4219 (2011) http://dx.doi.org/10.1016/j.ijhydene.2010.06.075.
5. A. S. Aricò, S. Srinivasan, and V. Antonucci, *Fuel Cells*, **1**, 133–161 (2001).
6. E. H. Yu, K. Scott, R. W. Reeve, L. Yang, and R. G. Allen, *Electrochim. Acta*, **49**, 2443–2452 (2004).
7. L. An, T. Zhao, X. Yan, X. Zhou, and P. Tan, *Sci. Bull.*, **60**, 55–64 (2015).
8. R. K. Raman and A. K. Shukla, *Fuel Cells*, **7**, 225–231 (2007).
9. G. Braesch et al., *J. Mater. Chem. A*, **8** (2020).

10. Z. Wang, J. Parrondo, C. He, S. Sankarasubramanian, and V. Ramani, *Nat. Energy*, **4** (2019).
11. Z. Wang, S. Sankarasubramanian, and V. Ramani, *Cell Reports Phys. Sci.*, **1** (2020).
12. Z. Wang et al., *ACS Appl. Energy Mater.* (2020).

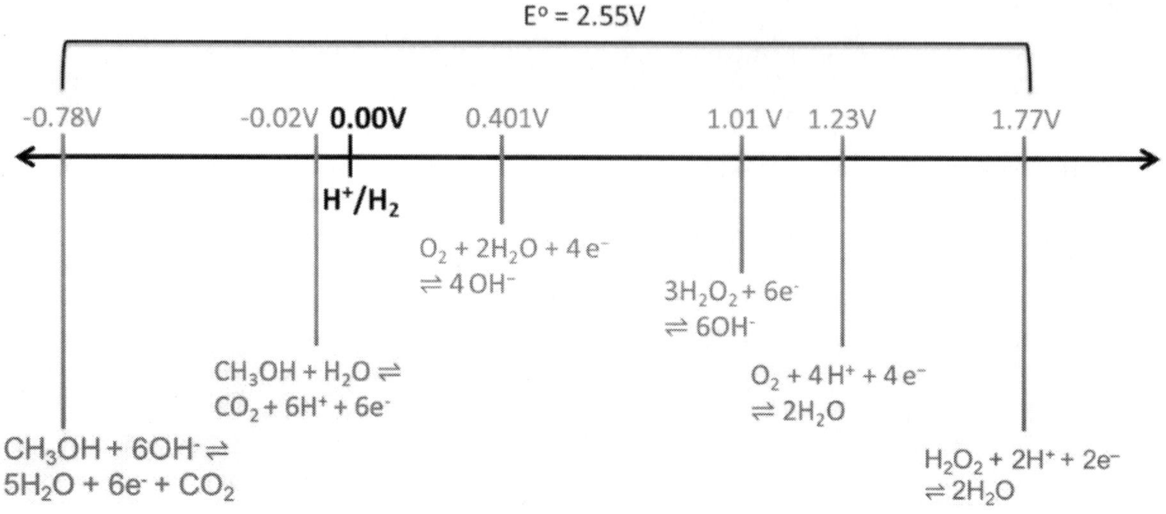

Figure 1. Theoretical potential for Methanol oxidation, oxygen reduction and hydrogen peroxide reduction in acidic and alkaline medium. Highest theoretical voltage is obtained with MOR in alkaline media combined with HPRR in acidic medium.

123c Intensifying Thermocatalytic Methanol Synthesis Using Electrochemical H_2 Activation

Samantha Roenigk, Evan Miu, Manjodh Kaur, James R. McKone

Abstract:

The alarming increase in atmospheric CO_2 concentration over the past century has driven significant global efforts to decarbonize chemical manufacturing through electrification and carbon reuse. The hydrogenation of CO_2 to valuable platform chemicals like methanol has gained significant attention, owing to the use of methanol as a raw material for numerous applications in chemical manufacturing. Advances in water electrolysis and renewable energy generation lend well to the use of electrolytic green hydrogen to convert CO_2 to methanol via established thermochemical processes.[1] Our lab is exploring the possibility of making further use of renewable electricity to intensify thermocatalytic methanol synthesis.

This presentation will discuss our work to build reactors that use electrochemical activation to drive methanol synthesis from syngas and CO_2/hydrogen mixtures under milder conditions (i.e., lower temperature and pressure) than can be achieved using thermal activation alone. Our approach involves using membrane electrode assemblies, analogous to those found in hydrogen fuel cells, to activate and deliver protons and electrons to a CO/CO_2 hydrogenation catalyst under a potential bias. We will discuss the construction and operation of these reactors along with studies directed at the hypothesis that conventional Cu-based syngas-methanol catalysts remain highly effective in this reactor configuration.[2] We will also discuss work to explore oxide-based catalysts that may benefit from the ability to accept reactive hydrogen as protons and electrons without the need to break H-H bonds. [3]

1. Jiang, Xiao, et al. 'Recent advances in carbon dioxide hydrogenation to methanol via heterogeneous catalysis.' *Chemical Reviews*15 (2020): 7984-8034.
2. Zhang, Sheng, et al. 'CO2 reduction: from homogeneous to heterogeneous electrocatalysis.' *Accounts of Chemical Research*1 (2020): 255-264.
3. Martin, Oliver, et al. 'Indium oxide as a superior catalyst for methanol synthesis by CO2 hydrogenation.' *Angewandte Chemie International Edition*21 (2016): 6261-6265.

123d Separation of diluted ethanol produced by CO_2 electrolysis can be cost effective

Magda Barecka, Joel W. Ager III, Alexei A. Lapkin

Electrolysis of CO_2 to ethanol, powered by renewable energy, is a promising pathway for a sustainable production of a molecule that can be used both as a precursor for chemical synthesis and as an energy source. However, a key challenge in the realization of the process in the industrial environment is the low concentration of ethanol streams obtained from electrocatalytic reactors (~1 to 10 vol%). As a result, required separation processes are believed to limit the economic viability and the possibilities of the scale-up of the process. Here we report how to overcome this bottleneck and present an experimental report of a tandem electrocatalytic reactor – porous membrane separator that allows to produce a concentrated ethanol stream (up to 40 vol%). By exploring the synergies between the optimal environment necessary for CO_2 electrolysis and subsequent separations, we present an overall process that allows to manufacture ethanol at a cost below the current market benchmark (0.9 USD/l).

123e Model-Based Development of a Novel Apparatus for Electrochemically Induced Crystallization of (di-)Carboxylic Acids

Jonas Goertz, Kristina Mielke, Marcel Gausmann, Andreas Jupke

Background:

Bio-based platform chemicals, such as carboxylic, amino or benzoic acids present a promising alternative to petrochemical synthesis routes. However, the downstream processing is still demanding and accounts for up to 60% of the total production costs despite intensive research [1]. The target molecules are biotechnologically produced at neutral pH and mostly present in the carboxylate form after fermentation. For separation by e.g. reactive extraction, crystallization or precipitation, multiple adjustments of the pH value are necessary. For this purpose, acids and bases are used, which produce neutral salts as by-products, reduce the economic efficiency and increase the carbon footprint of bio-based platform chemicals.

Method:

This contribution presents the model-based development of a novel electrochemical cell for the pH shift crystallization of carboxylic acids. Instead of additives, the adjustment of the pH, which triggers the crystallization of the acid, is conducted by employing water-splitting electrolysis [2-4]. The scale-up of the electrochemical crystallization requires novel electrolysis cells that operate reliable even when dispersed solid crystals are present in the electrolyte. We applied Computational Fluid Dynamics (CFD) to design suitable geometries for the electrolyte chambers, which can cope with liquid-gas as well as liquid-gas-solid containing electrolytes. Especially the diameter of the bubbles strongly influences the gas-liquid interaction, but literature regarding the diameter of hydrogen and oxygen gas bubbles inside electrolyser with small gaps is sparse [5]. Therefore, we experimentally investigated the influence of operating parameters on the diameter of oxygen bubbles in the anodic chamber of a $100cm^2$ electrolysis cell. Using a high-speed camera with a transparent electrolysis design and a neural network for bubble recognition, the effect of different current densities, volumetric flow rates and electrolyte concentrations on the bubble diameter is quantified [6].

Discussion andamp; Conclusion:

Preliminary simulation and literature [5] indicates that the fluid pattern is governed by the nucleation and induced convection of the rising gas bubbles. Based on these findings, a three-phase (gas-solid-liquid) Euler-Lagrange simulation algorithm was created inside the OpenFOAM environment to describe the effect of the disperse phases on the continuous phase. With this approach different prototypes for electrolysis cells were developed, 3D-printed and investigated in the laboratory with respect to their hydrodynamic behavior. The simulation results are validated with the experimentally measured residence time distributions and predicted the increase of Bodenstein number due to the bubble generation. The information gained from the experiments and simulation is used in a model aided feedback loop to generate geometric designs with improved performance regarding the fluid, gas and particle residence time. The designs are subsequently tested by rapid prototyping. Different concepts for gas-liquid flows like co-current or crosscurrent are presented and their advantages and disadvantages are discussed. With the developed CFD model and the rapid prototyping aided workflow we were able to design cell geometries tailored for the electrochemical crystallization. The new designs achieved stable operation with particles present inside the electrolysis cell. Besides, a classification according to particle size could be observed, when using a crosscurrent approach. While large crystals were successfully withdrawn from the suspension outlet at the bottom of the chamber, fine particles exited with the liquid. This enables the automatic separation of product and the recovery of fine particles, which can then be returned to the electrolysis cell with fresh feed solution and be used as seeds. We envision the new modules developed for electrochemically induced crystallization to overcome the waste burden and mitigate the environmental footprint of industrial biotechnology by providing a technology for waste free recovery of succinic acid.

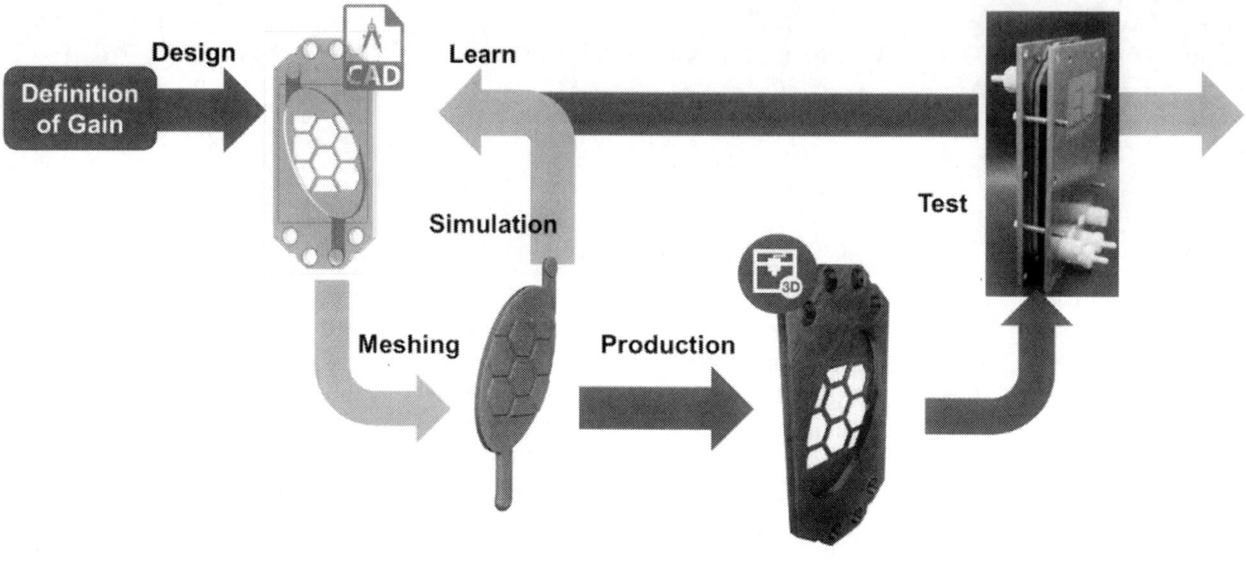
Definition of Gain
Design
CAD
Learn
Simulation
Meshing
Production
3D
Test

Investigation of Current Production and Metal Removal from Synthetic Desalter Effluent Using a Dual Microbial Fuel Cell with *Shewanella oneidensis* MR-1

Carlos Munoz-Cupa[1], Amarjeet Bassi[1]
[1]Department of Chemical and Biochemical Engineering, The University of Western Ontario, London, Ontario Canada
Corresponding author: Department of Chemical and Biochemical Engineering, Thompson Engineering Building, Room TEB 467, Western University Tel: 519-661-2111 ext. 88324 / 88219 Fax: 519-661-3498, abassi@uwo.ca

Abstract

Desalter effluent treatment has multiple challenges such as chemical processing after treatment and expensive methods. Due to these disadvantages, novel studies for heavy metal removal from desalter effluent have gained popularity. This research focused on heavy metal removal and current production from synthetic desalter effluent using a dual-chamber Microbial Fuel Cell (MFC) inoculated with *Shewanella oneidensis* MR-1 in the anode. *S. oneidensis* MR-1 was used due to its ability for heavy metal reducing in their extracellular wall without affecting the electron transport from the nutrient source to the electrode. The catholyte solution was potassium hexacyanoferrate III ($K_3Fe(CN_6)$) dissolved in 100 mM phosphate buffer solution and it was separated from the anode by a cation exchange membrane and connected to the anode with an external resistance of 1000 Ω using carbon felt electrodes with an area of 0.0036 m2. The synthetic desalter effluent was composed by $ZnSO_47H2O$ (13.5 mg/L), $MgSO_4$ (415 mg/L), $CuSO_45H2O$ (10.5 mg/L), $MnSO_4H2O$ (7.5 mg/L), NaCl (2410 mg/L) and phenol (354.2 mg/L), and the total metals in solution were evaluated by ICP-OES. Synthetic desalter effluent was inoculated with 52.5 mL of inoculum at OD600 of 0.25. The nutrient source composed of tryptic soy broth and M9 mineral solution in 18 mM was fed into the anode every 24 hours, and catholyte solution was fed into the cathode every 48 hours. The MFC was operated for 140 hours with these conditions. Additionally, biofilm formation on the electrode was evaluated by scanning electron microscopy (SEM). The maximum current density and open-circuit voltage (OCV) were 70.3 mA/m2 and 252.9 mV, respectively. The heavy metal removal showed a Cu (II), Mg (II), Mn (II), and Zn (II) removal of 11%, 65%, 79%, and 60%, respectively. SEM analysis of the anode electrode showed formation of bacterial nanowires that enhanced the electron transport from the nutrient source to the cell and the electrode surface. With these results, MFCs could arise as a new method for heavy metal removal and energy production.

Keywords: Microbial fuel cell, heavy metal, bacterial nanowires, *Shewanella oneidensis*, wastewater

INTRODUCTION

Shewanella oneidensis MR-1 is a metal reducing bacteria that have increased its research impact due to its metabolic pathways. These pathways use the intracellular reductases for reduction of compounds like heavy metals.[1] Furthermore, desalter units produce large quantities of desalter effluent in the washing process for pollutant removals from crude oil. Approximately 3200 bbl/day are produced per 1000 bbl/day of crude oil treated. Conventional methods such as precipitation, absorption, or membrane technologies have some disadvantages like high cost and low efficiency under low heavy metals concentrations.[2,3] New technologies have been under intensive research for solving these and other problems related to chemical disposal.[4] Microbial Fuel Cells (MFC) have gaining importance due to their ability to generate electricity while removing pollutants. The process uses the oxidation and reduction of carbon sources and heavy metal ions producing voltage in response to the electron transport from anode to cathode.[5,6] The transport is facilitated by electrodes and external resistance via an external circuit. Additionally, the protons go through a Proton Exchange Membrane (PEM) from the anode to the cathode for the reduction in the cathode chamber.[7,8]

Researches have investigated removal of heavy metals such as Mn (IV) and Fe (III) using *S. oneidensis* with dual chamber MFC.[9] Moreover, catholyte reduction and precipitation of Fe (III), Cu (II), and Zn (II) using MFC were investigated using mixed cultures in the anode.[10] Additionally, Cr (VI) reduction in the biotic cathode with *S. oneidensis*,[11,12] and in an abiotic cathode[13] with a dual MFC having removal efficiencies over 80% for both cases were studied. Other MFC configurations using hybrid electrolysis cell systems were used [14] for Cr (VI), Pb (II), and Ni (II) removal. In addition, soil heavy metals using a MFC was studied using Cr (VI), Cu (II) and Pb (II).[7,15,16] MFC shows potential for power generation and oxidation/reduction of heavy metals simultaneously; this characteristic enhances the use of the technology for wastewater treatment and power devices such as sensors. However, in some cases, the high reduction potential of Pb, Cd, and Ni is the main disadvantage in catholyte reduction overcoming the power generation in the MFC.[17]

In this study, *Shewanella oneidensis* MR-1 was evaluated in a dual-chamber MFC for heavy metal removal and voltage production from synthetic desalter effluent composed of Cu (II), Zn (II), Mg (II), Mn (II), Na and Phenol. The results showed that *S. oneidensis* is a promising bacterium for heavy metal removal from salty wastewaters with simultaneous power generation.

MATERIALS AND METHODS

Synthetic desalter effluent was studied according to 5 different concentrations (**Table 1**). Additionally, potassium hexacyanoferrate III ($K_3Fe(CN_6)$) was used as catholyte solution. Furthermore, *S. oneidensis* was inoculated with tryptic soy broth (TSB) at 30°C for 24 h before taking a partial of this liquid culture for inoculation of the anode chamber in the MFC. We used a dual chamber MFC separated by a CMI-7000A membrane of 3.3 cm diameter and with carbon felt electrodes (6 cm x 6 cm x 0.8 cm) positioned in each chamber. These electrodes were connected through an external resistance (1kΩ) by copper wires.

Table 1. Synthetic wastewater composition used for the MFC experiments.

Compound	Conc. 1	Conc. 2	Conc. 3	Conc. 4	Conc. 5
$ZnSO_4 \cdot 7H_2O$	13.5 mg/L	27.0 mg/L	40.5 mg/L	500 mg/L	13.5 mg/L
$MnSO_4H_2O$	7.5 mg/L	15.0 mg/L	22.5 mg/L	500 mg/L	7.5 mg/L
$MgSO_4$	415.0 mg/L	830.0 mg/L	1245.0 mg/L	500 mg/L	415.0 mg/L
$CuSO_4 \cdot 5H_2O$	10.5 mg/L	21.0 mg/L	31.5 mg/L	500 mg/L	10.5 mg/L
NaCl	2410.0 mg/L	2410.0 mg/L	2410.0 mg/L	500 mg/L	2410.0 mg/L
C_6H_5OH	NA	NA	NA	NA	

RESULTS AND DISCUSSION

Biofilm formation was observed as a single layer on the electrode surface after 7 days. The carbon felt fibers on the electrode facilitate the attachment of bacteria to the surface (**Figure 1a**). The external resistance (1 kΩ) contributes to the formation of thin biofilm. However, biofilm showed a higher concentration (**Figure 1b**) in comparison with the biofilm formation observed by other authors.[18] Additionally, biofilm acts as the mediator for easy electron transportation from the medium to the electrode, and the electrons are taken directly from the biofilm.[19]

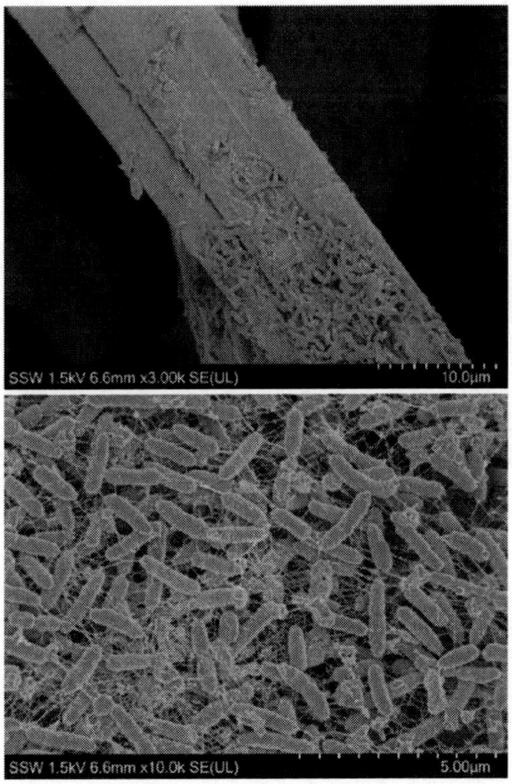

a)

b)

Figure 1. ***Shewanella oneidensis*** **biofilm attached to carbon felt fibers in the bio-anode. a) Carbon felt fibers with biofilm, b)** ***S. oneidensis*** **biofilm with nanowires**

Bacterial nanowires demonstrate the connectivity between cells increasing electron transport between the medium and the electrode (**Figure 1b**). *S. oneidensis* produce nanowires with biofilm formation when added even to a mixed culture with an o-xylene medium,[20] implying the adaptation of the bacteria to adverse conditions. Additionally, microbial nanowires increase the resistance to metal toxicity and might help in metal reduction due to the increase in the surface area.[21] *S. oneidensis* nanowires are an extension of periplasmic membrane, with a similar composition to the cell membrane, leading to a higher electron transport with high conductivity (0.01 – 1 S/cm).[22] This characteristic assists the electron transport mechanism, that direct transfer electrons to the electrode through the outer cell membrane.[23] Bacterial nanowires act as response of adverse conditions with high toxic compounds such as heavy metals.

Power density curves have trends that showed the maximum power density that every concentration in the anode and cathode produced at different external resistances (**Figure 2**). The maximum power density of 76766.7 mW/m^2 at a current density of 326.5 mA/m^2 was reached with Conc. 5 in the anode chamber (**Figure 2a**). High metal concentration significantly reduces the power density, which contribute to a reduction in electron transport due to high deposition in the bacteria surface or biosorption. These results indicate that metals affect the MFC performance due to their toxic inhibition and not the electron transport in the anode.[24]

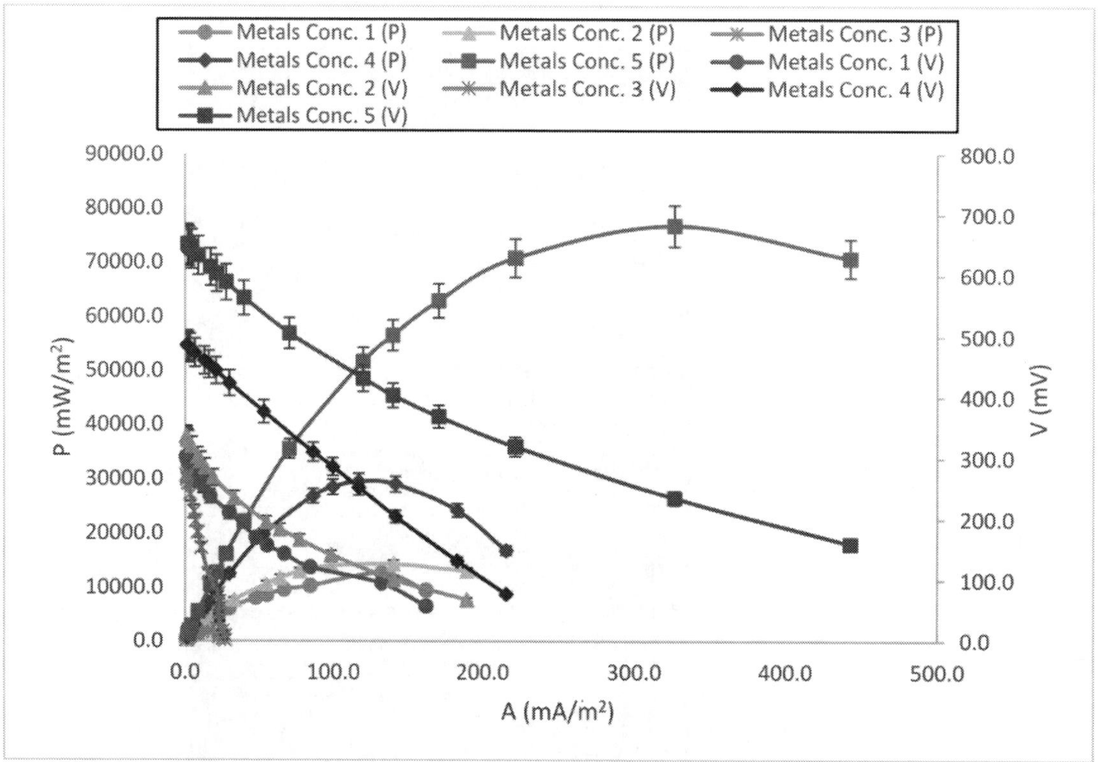

a)

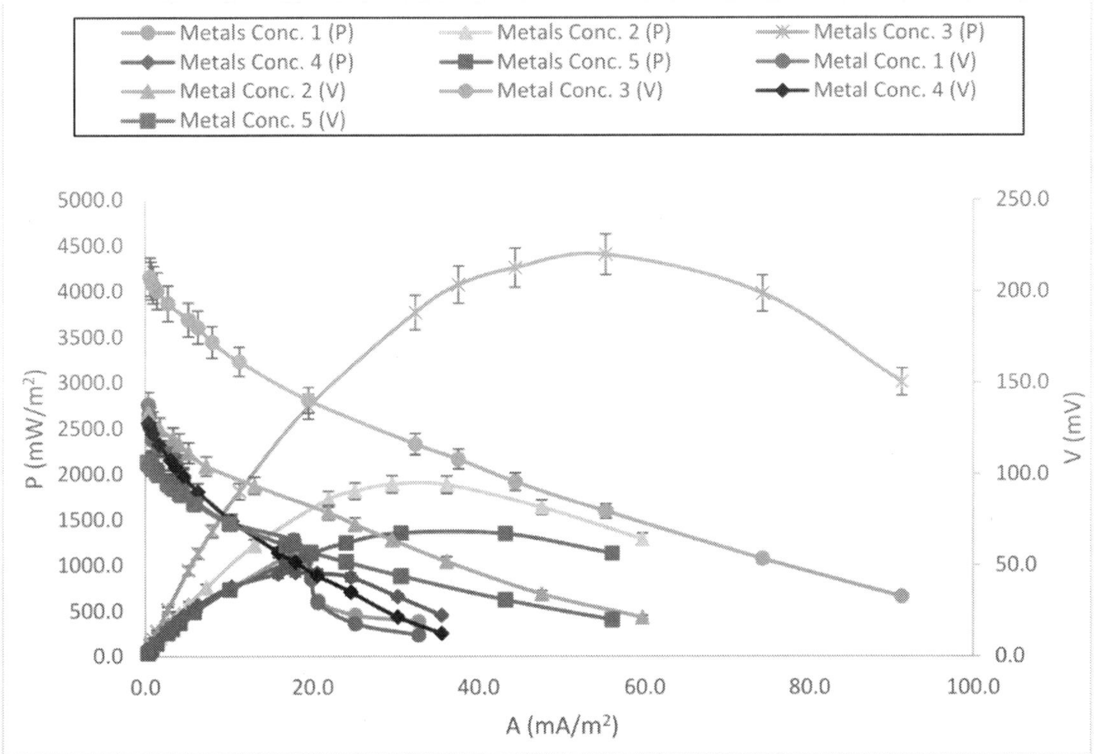

b)

Figure 2. Polarization curves in bio-anode and abiotic cathode varying external resistance from 100 Ω to 100000 Ω. a) Metal concentrations in bio-anode, and b) metal concentrations in abiotic cathode.

Power density curves for metal solutions in the cathode without the catholyte solution presented a similar behavior in comparison with metals in the anode chamber. A maximum power density of 4400.1 mW/m^2 at a current density of 55.3 mA/m^2 was reached with Conc. 3 in the cathode chamber (**Figure 2b**). Additionally, $K_3Fe(CN_6)$ stabilized the voltage at higher resistances due to the higher redox potential of this salt in comparison with the metals in the cathode (**Figure 2b**). Activation losses at low R_{ext} in the initial stage of the polarization curves showed a decrease in the voltage, then at higher R_{ext} over 800 Ω, the voltage is further decreasing due to ohmic losses represented as the most linear part of the polarization curves, leading to a movement of electrons from substrates through the circuit. Therefore, concentration losses occurred due to the lack of organic sources for bacterial metabolism.[25] Furthermore, the high internal resistances produce less power density because of greater destabilization.[7] One possible solution to improve internal resistance is to reduce distance between electrodes,[15] achieving higher power production.

The maximum OCV in the anode was at the beginning of the experiment for Conc. 5, Conc. 4, and Conc. 1 with results of 517.55 mV, 243.20 mV, and 517.55 mV, respectively (**Figure 3a**). At t = 50 h and t = 96 h, maximum OCV for Conc. 2 and 3 were 254.40 mV and 135.75 mV, respectively. Metal concentration might be toxic for the bacteria growth, and the stabilization of the bacteria reached after 72 h in the absence of metals and phenol. In these optimal conditions *S. oneidensis* accelerates the electron transfer reducing the resistance losses and increasing the output voltage.[26–28] Additionally, the voltage decreased for all the metal concentrations and stabilization was reached after 72 h of operation, which in this case was the MFC stabilization.

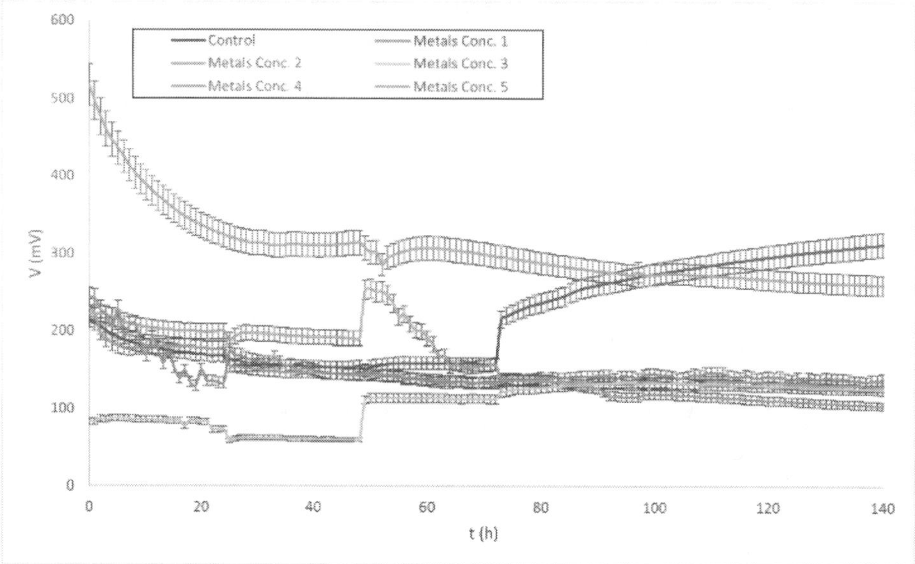

a)

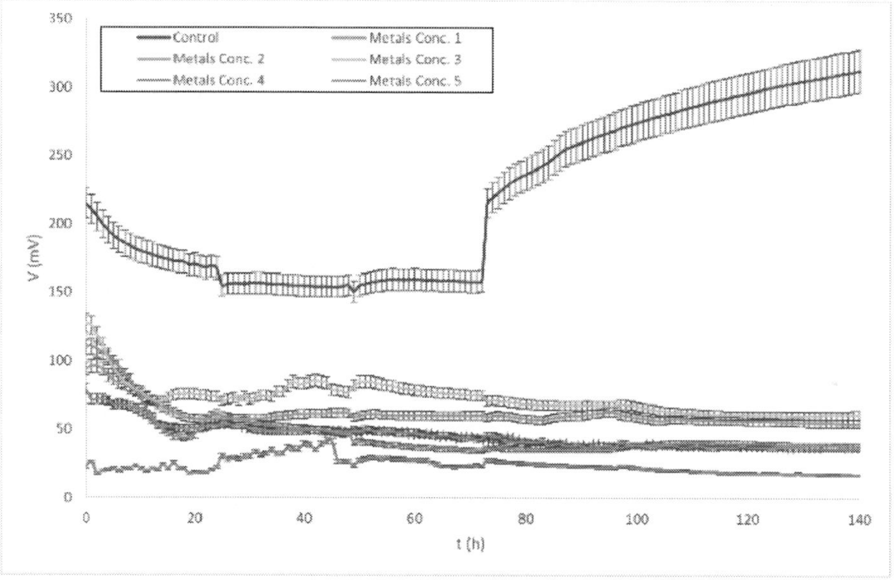

b)

Figure 3. Output Current Voltage (OCV) for the duration of 140 h of MFC operation. a) Metals in the bio-anode, and b) metals in the abiotic cathode.

Metals in the cathode showed a lower voltage production in comparison with the control experiments (**Figure 3b**). The maximum OCV was 127.70 mV for Conc. 3, and the minimum was 17.09 for Conc. 4. After 72 h a constant trend in the voltage produced was observed, similar to the results reported with metals in the anode chamber.

Heavy metal removal from synthetic wastewater was evaluated in the anode and cathode chambers. *S. oneidensis* facilitated metal reduction with the highest concentration removal of Cu (II) which was 93.00% at Conc. 4, Mg (II) of 96.94% at Conc. 3, Mn (II) of 97.62% at Conc. 3,

Zn (II) of 95.80% at Conc. 4, and Na (I) of 61.19% at Conc. 4. Furthermore, synthetic desalter effluent achieved metal removals greater than 84.71% in the anode (**Figure 4**). However, the initial concentration of sodium was high, which resulted in a low removal of 35.82%. *S. oneidensis* respiration pathway enhances electron transport because of the production of riboflavin that reduces metals simultaneously, as consequence metals can also be deposited on the electrode surface.[15] Single cultures like *S. oneidensis* enhances metal removal, due to their metal reducing characteristics in the extracellular membrane with the cytochromes MtrC and OmcA that act as metal terminals for the extracellular reduction.[29,30]

Metal removal in the cathode chamber reached a maximum removal for Cu (II) at 99.78% at Con. 4, Mg (II) at 63.86% at Con. 3, Mn (II) at 58.67% at Conc. 2, Zn (II) 60.74% at Conc. 2 and Na (I) 68.66% at Conc. 5. The higher redox potential of Cu (II) (+0.34 V) indicates that the reduction of this compound took more electrons for reduction over the other metals, observing higher removals of Cu over Mg, Mn, and Zn.

CONCLUSIONS

S. oneidensis is a metal reducing bacteria able to reduce heavy metals at low concentrations facilitating the electron transport from the cell surface to electrodes increasing the viability for power production in MFC. Furthermore, bacterial nanowires increase the biofilm attachment and enhance electron transport increasing the power production, the output voltage, and the resistance to toxic compounds such as heavy metals. Additionally, dual MFC configuration have the advantage of treating metal solutions simultaneously in the anode and the cathode making this set-up convenient for wastewater treatment. However, studies related to electrode size, distance between electrodes and membrane size are required to increase the efficiency and the viability for cost-effective applications.

ACKNOWLEDGEMENT

The authors acknowledge the financial support from Natural Sciences and Engineering Research Council of Canada (NSERC) through the Discovery Grants.

REFERENCES

1. Yin Y, Liu C, Zhao G, Chen Y. Versatile mechanisms and enhanced strategies of pollutants removal mediated by Shewanella oneidensis: A review. *J Hazard Mater*. 2022;440:129703. doi:10.1016/j.jhazmat.2022.129703

2. Sun DT, Peng L, Reeder WS, Moosavi SM, Tiana D, Britt DK, Oveisi E, Queen WL. Rapid, Selective Heavy Metal Removal from Water by a Metal-Organic Framework/Polydopamine Composite. *ACS Cent Sci*. 2018;4(3):349-356. doi:10.1021/acscentsci.7b00605

3. Ezziat L, Elabed A, Ibnsouda S, El Abed S. Challenges of microbial fuel cell architecture on heavy metal recovery and removal from wastewater. *Front Energy Res.* 2019;7(January):1-13. doi:10.3389/fenrg.2019.00001

4. Kaushik A, Singh A. Metal removal and recovery using bioelectrochemical technology: The major determinants and opportunities for synchronic wastewater treatment and energy production. *J Environ Manage.* 2020;270(May):110826. doi:10.1016/j.jenvman.2020.110826

5. Logan BE. *Microbial Fuel Cells*. Wiley-Inte. John Wiley and Sons Inc.; 2008.

6. Munoz-Cupa C, Hu Y, Xu C, Bassi A. An overview of microbial fuel cell usage in wastewater treatment, resource recovery and energy production. *Science of the Total Environment.* 2021;754:142429. doi:10.1016/j.scitotenv.2020.142429

7. Abbas SZ, Rafatullah M, Ismail N, Shakoori FR. Electrochemistry and microbiology of microbial fuel cells treating marine sediments polluted with heavy metals. *RSC Adv.* 2018;8(34):18800-18813. doi:10.1039/C8RA01711E

8. al Shahrani AA. *Metal Ions Impact on Shewanella Oneidensis MR-1 Adhesion to ITO Electrode and the Enhancement of Current Output*. Loyola University Chicago; 2020.

9. Bretschger O, Obraztsova A, Sturm CA, In SC, Gorby YA, Reed SB, Culley DE, Reardon CL, Barua S, Romine MF, Zhou J, Beliaev AS, Bouhenni R, Saffarini D, Mansfeld F, Kim BH, Fredrickson JK, Nealson KH. Current production and metal oxide reduction by Shewanella oneidensis MR-1 wild type and mutants. *Appl Environ Microbiol.* 2007;73(21):7003-7012. doi:10.1128/AEM.01087-07

10. Gajda I, Stinchcombe A, Greenman J, Melhuish C, Ieropoulos I. Microbial fuel cell – A novel self-powered wastewater electrolyser for electrocoagulation of heavy metals. *Int J Hydrogen Energy.* 2017;42(3):1813-1819. doi:10.1016/j.ijhydene.2016.06.161

11. Hsu L, Masuda SA, Nealson KH, Pirbazari M. Evaluation of microbial fuel cell Shewanella biocathodes for treatment of chromate contamination. *RSC Adv.* 2012;2(13):5844-5855. doi:10.1039/c2ra20478a

12. Xafenias N, Zhang Y, Banks CJ. Enhanced performance of hexavalent chromium reducing cathodes in the presence of Shewanella oneidensis MR-1 and lactate. *Environ Sci Technol.* 2013;47(9):4512-4520. doi:10.1021/es304606u

13. Li M, Zhou S, Xu Y, Liu Z, Ma F, Zhi L, Zhou X. Simultaneous Cr(VI) reduction and bioelectricity generation in a dual chamber microbial fuel cell. *Chemical Engineering Journal.* 2018;334(October 2017):1621-1629. doi:10.1016/j.cej.2017.11.144

14. Li Y, Wu Y, Liu B, Luan H, Vadas T, Guo W, Ding J, Li B. Self-sustained reduction of multiple metals in a microbial fuel cell-microbial electrolysis cell hybrid system. *Bioresour Technol.* 2015;192:238-246. doi:10.1016/j.biortech.2015.05.030

15. Abbas SZ, Rafatullah M, Ismail N, Syakir MI. A review on sediment microbial fuel cells as a new source of sustainable energy and heavy metal remediation: mechanisms and future prospective. *Int J Energy Res*. 2017;41(9):1242-1264. doi:10.1002/er.3706

16. Habibul N, Hu Y, Sheng GP. Microbial fuel cell driving electrokinetic remediation of toxic metal contaminated soils. *J Hazard Mater*. 2016;318:9-14. doi:10.1016/j.jhazmat.2016.06.041

17. Chakraborty I, Sathe SM, Khuman CN, Ghangrekar MM. Bioelectrochemically powered remediation of xenobiotic compounds and heavy metal toxicity using microbial fuel cell and microbial electrolysis cell. *Mater Sci Energy Technol*. 2020;3:104-115. doi:10.1016/j.mset.2019.09.011

18. Liu T, Yu YY, Li D, Song H, Yan X, Chen WN. The effect of external resistance on biofilm formation and internal resistance in Shewanella inoculated microbial fuel cells. *RSC Adv*. 2016;6(24):20317-20323. doi:10.1039/c5ra26125b

19. Vanegas-Hernández DM, Cardona-Aristizabal ML, Zapata-Benabithe Z. Evaluation of Activated Carbon Electrodes as Anodes in a Microbial Fuel Cell Using Shewanella Putrefaciens. *Revista Facultad de Ingenieria*. 2020;29(54). doi:10.19053/01211129.v29.n54.2020.10468

20. You J, Deng Y, Chen H, Ye J, Zhang S, Zhao J. Enhancement of gaseous o-xylene degradation in a microbial fuel cell by adding Shewanella oneidensis MR-1. *Chemosphere*. 2020;252:126571. doi:10.1016/j.chemosphere.2020.126571

21. Sure S, Ackland ML, Torriero AAJ, Adholeya A, Kochar M. Microbial nanowires: An electrifying tale. *Microbiology (United Kingdom)*. 2016;162(12):2017-2028. doi:10.1099/mic.0.000382

22. Creasey RCG, Mostert AB, Nguyen TAH, Virdis B, Freguia S, Laycock B. Microbial nanowires – Electron transport and the role of synthetic analogues. *Acta Biomater*. 2018;69:1-30. doi:10.1016/j.actbio.2018.01.007

23. Xu L, Zhao Y, Owusu KA, Zhuang Z, Liu Q, Wang Z, Li Z, Mai L. Recent Advances in Nanowire-Biosystem Interfaces: From Chemical Conversion, Energy Production to Electrophysiology. *Chem*. 2018;4(7):1538-1559. doi:10.1016/j.chempr.2018.04.004

24. Mathuriya AS, Yakhmi J v. Microbial fuel cells to recover heavy metals. *Environ Chem Lett*. 2014;12(4):483-494. doi:10.1007/s10311-014-0474-2

25. Bhunia P, Dutta K. Biochemistry and electrochemistry at the electrodes of microbial fuel cells. In: *Progress and Recent Trends in Microbial Fuel Cells*. Elsevier; 2018:327-345. doi:10.1016/B978-0-444-64017-8.00016-6

26. Dai HN, Duong Nguyen TA, My LE LP, Tran M van, Lan TH, Wang CT. Power generation of Shewanella oneidensis MR-1 microbial fuel cells in bamboo fermentation effluent. *Int J Hydrogen Energy*. 2021;46(31):16612-16621. doi:10.1016/j.ijhydene.2020.09.264

27. Park YH, Smith G, Park E. Mediator-less microbial fuel cell employing Shewanella Oneidensis. *Energy Sources, Part A: Recovery, Utilization and Environmental Effects*. 2016;38(12):1779-1784. doi:10.1080/15567036.2014.969815

28. Zhang P, Liu J, Qu Y, Feng Y. Enhanced Shewanella oneidensis MR-1 anode performance by adding fumarate in microbial fuel cell. *Chemical Engineering Journal*. 2017;328:697-702. doi:10.1016/j.cej.2017.07.008

29. Gude VG. Wastewater treatment in microbial fuel cells - An overview. *J Clean Prod*. 2016;122:287-307. doi:10.1016/j.jclepro.2016.02.022

30. Han JC, Chen GJ, Qin LP, Mu Y. Metal Respiratory Pathway-Independent Cr Isotope Fractionation during Cr(VI) Reduction by Shewanella oneidensis MR-1. *Environ Sci Technol Lett*. 2017;4(11):500-504. doi:10.1021/acs.estlett.7b00471

635a First European Demonstration Plant for the Synthesis of Renewable Ome Fuels

Alvaro Ferre, Johannes Voggenreiter, Christian F. Breitkreuz, Denis Worch, Udo Lubenau, Hans Hasse, Jakob Burger

Poly(oxymethylene) dimethyl ethers (OME) are oligomers of the structure CH_3-O-$[CH_2O]_n$-CH_3. OME of a chain length of 3 to 5 (OME_{3-5}) are particularly interesting, because they have diesel-like properties. It has been demonstrated that OME_{3-5} significantly reduce soot formation in engine test stands [1] and under real driving conditions [2]. OME_{3-5} are produced in all industrially feasible processes exclusively from the raw material methanol, which presents a flexible raw material base, including biomass, carbon dioxide, other organic waste, and natural gas. Within the framework of the project consortium "NAMOSYN" funded by the German Federal Ministry of Education and Research (BMBF), the Technical University of Munich has erected Europe's first continuous production demonstration plant for OME synthesis with a capacity of 1 kg/h (= 8 tonnes/a) OME_{3-5} at the Campus Straubing [3]. Its process flow diagram is shown in Figure 1 [4]. The feed, a methanolic formaldehyde solution with about 10 mass-% water, is mixed with two liquid recycle streams and fed to the reactor. In the heterogeneously acid-catalyzed reactor, a stream comprising formaldehyde (FA), water (WA), methanol (ME), and OME of various chain lengths is obtained. The isolation of the desired product fraction OME_{3-5} is carried out in two distillation columns C1 and C2. The first distillation column C1 separates OME of chain lengths $n \geq 3$ as bottom product from a stream containing mainly FA, ME, WA and OME_{1-2}. The second distillation column C2 separates the product fraction OME_{3-5} as overhead product from OME of longer chain lengths $n \geq 6$ which are recycled back to the reactor. Before the distillate of the column C1 is recycled to the reactor, water is removed in a membrane unit.

The most challenging unit of the process is the reactive distillation column C1 because of the high number of components present and because formaldehyde reacts uncatalyzed with methanol and water. Further, formaldehyde may easily precipitate since its solubility in these mixtures is limited. An equilibrium stage model for the reactive column C1 was developed, which is illustrated in Figure 2. This model includes a novel solid-liquid equilibrium model for formaldehyde conceived by the University of Kaiserslautern [5]. For the first time, the column is verified experimentally at process-relevant conditions including high concentrations of formaldehyde in the feed (> 0.10 g/g).

A further challenge is the separation of water from mixtures containing formaldehyde, methanol and OME. Because of the presence of multiple azeotropes, distillation is not the preferred option for this task [4]. Therefore, our plant uses a membrane unit, developed by the German company DBI Gas- und Umwelttechnik GmbH, that contains an innovative inorganic membrane material to selectively remove water from formaldehyde-containing mixtures as it was also but on a smaller scale investigated by Schmitz et al. [6].

In this contribution, we introduce the demonstration plant for OME synthesis at the Campus Straubing of the Technical University of Munich. The main challenges regarding the operation of the different units are discussed. We present the results of long-term steady-state experiments of the plant running in closed loop. In detail, we are looking at the performances of the reactor, the distillation column C1, and the membrane unit. Experimental profiles are presented and compared to process simulations. The present work shows that the individual as well as the combined operation of all process units is feasible despite the existing challenges and presents a relevant step on the way to a non-fossil transportation era.

References

[1] M. Härtl, K. Gaukel, D. Pélerin, G. Wachtmeister. *MTZ Worldw.* 78 (2) (2017) 52-29. DOI: 10.1007/s38313-016-0163-6

[2] M. Münz, A. Mokros, D. Töpfer, C. Beidl. *MTZ Worldw.* 79 (3) (2018) 16-21. DOI: 10.1007/s38313-

017-0185-8

[3] A. Ferre, J. Voggenreiter, Y. Tönges, J. Burger. *MTZ Worldw.* 82 (5) (2020) 26-42. DOI: 10.1007/s38313-021-0639-x

[4] N. Schmitz, E. Ströfer, J. Burger, H. Hasse. *Ind. Eng. Chem. Res.* 56 (2017) 11519-11530. DOI: 10.1021/acs.iecr.7b02314

[5] C. F. Breitkreuz, J. Burger, H. Hasse, Ind. Eng. Chem. Res. 61 (2022) 1871-1884. DOI: 10.1021/acs.iecr.1c04275

[6] N. Schmitz, C. F. Breitkreuz, E. Ströfer, J. Burger, H. Hasse. J. Membr. Sci. 564 (2018) 806-812. DOI: 10.1016/j.memsci.2018.07.053

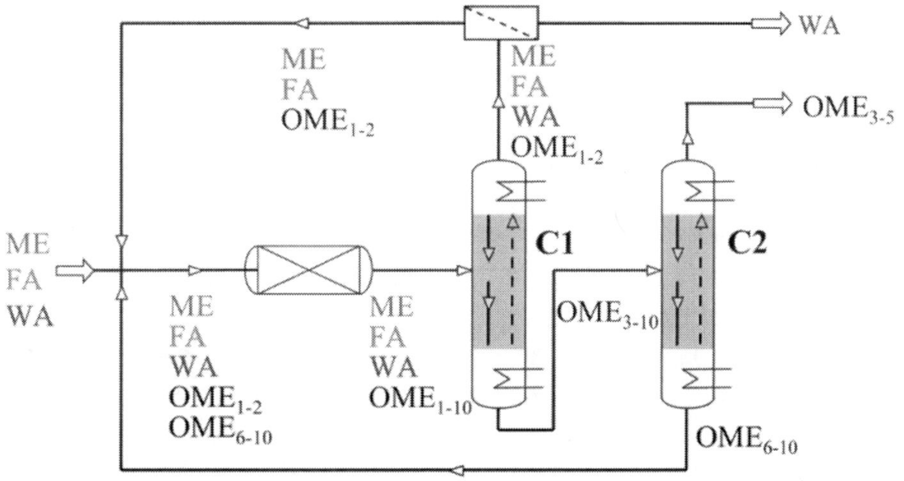

Fig. 1: Process flow diagram of the process to convert methanol and aqueous formaldehyde to $OME_{3\text{-}5}$ and water [4].

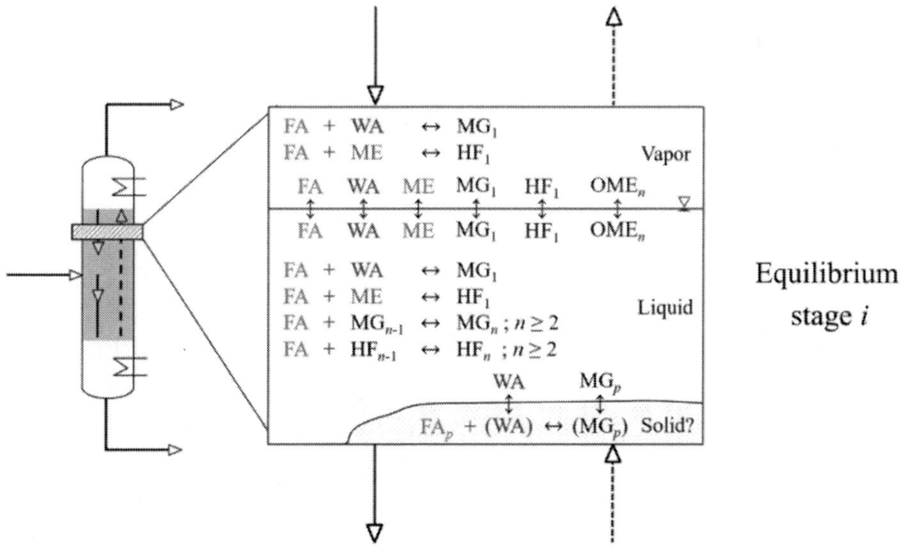

Fig. 2: Scheme of the reactive equilibrium stage model including the potential solid precipitation of formaldehyde used to model the distillation column C1 [5].

ADDITIVE SOLUTIONS FOR RENEWABLE FUELS AND FEEDSTOCKS

Paper 635b

Daniel Dreyer[1], Nestor Soriano[1], Ashish Dhawan[2], Oussama Zenasni[1], Josep Marca[1], Elizabeth Balapitiya[1]

1. Nalco Water, An Ecolab Company; Sugar Land, TX
2. Ecolab; Naperville, IL

Abstract

In this paper, we will discuss experimental findings regarding the chemical and physical properties of biofuels and their feedstocks, differences between biofuels/feedstocks and their petroleum-based counterparts, and a variety of chemical treatments that can be used to alter the properties of biofuels and feedstocks. Treatment requirements vary significantly depending on the biomass feedstock, as well as the process conditions (e.g., transesterification versus hydrotreatment). Our goal herein is to provide a broad overview of these differences, as well as an overview of chemical treatments that can be used to ensure that key specifications are met. In some cases, those specifications are prescriptively defined (e.g., conductivity and lubricity requirements outlined in ASTM D975), but in other cases the specifications are subjective and left to individual producers or end users (e.g., cold flow requirements for feedstocks). Key areas that will be discussed in this paper include cold flow improvers (e.g., pour point depressants, cloud point depressants, CFPP depressants), conductivity improvers (also known as anti-static additives), lubricity improvers, and antioxidants.

Introduction

Increasing pressure from regulatory authorities as well as end-use consumers is driving refiners and fuel producers away from petroleum-derived hydrocarbons and toward more sustainable hydrocarbons that can support a circular economy.[1-3] A broad mixture of alternative energy sources continues to be developed and commercialized (e.g., electric vehicles, solar panels, wind turbines, etc.), but combustible fuels derived from biological hydrocarbon sources will continue to be a key piece of that energy mixture.[4] Some types of biofuels such as biodiesel (e.g., FAMEs; fatty acid methyl esters) have been known and commercially produced for decades. Other forms of biofuels such as hydrotreated vegetable oil (HVO), also referred to as renewable diesel (RD), are somewhat newer, particularly in US-based markets, but have likewise been produced for many years (see Figure 1 for a comparison of these fuels).

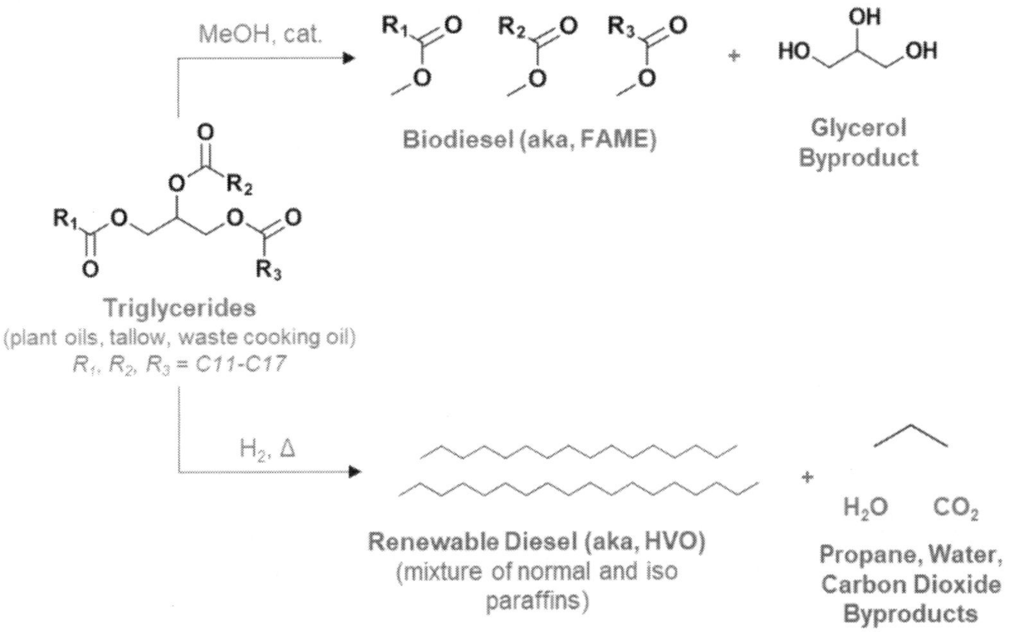

Figure 1. Reaction pathways for the production of biodiesel (top) or renewable diesel (bottom) from a common, bio-based feedstock.

Despite their long history, both actual and planned production of biofuels is increasing substantially, and the relative mixture of fuels being produced is also expected to undergo a significant shift in the near term. The Energy Information Agency (EIA) has recently reported that RD production has been growing (see Figure 2), is projected to surpass biodiesel production for the first time ever in 2022, and continue to increase thereafter.[5] Concurrently, conventional ULSD production is projected to decrease as fuel producers shift away from petroleum sources to meet increasing demand for renewable fuels. In broad terms, these dynamics are widely acknowledged within the industry, but appear to be gaining momentum as highlighted by the EIA report. While treatment strategies for petroleum-derived fuels have been studied and perfected for decades, the requirements and preferred solutions for biofuels are less well understood. This gap needs to be addressed in order to ensure that the rapid increase in biofuel production and use is accomplished in a safe and reliable manner.

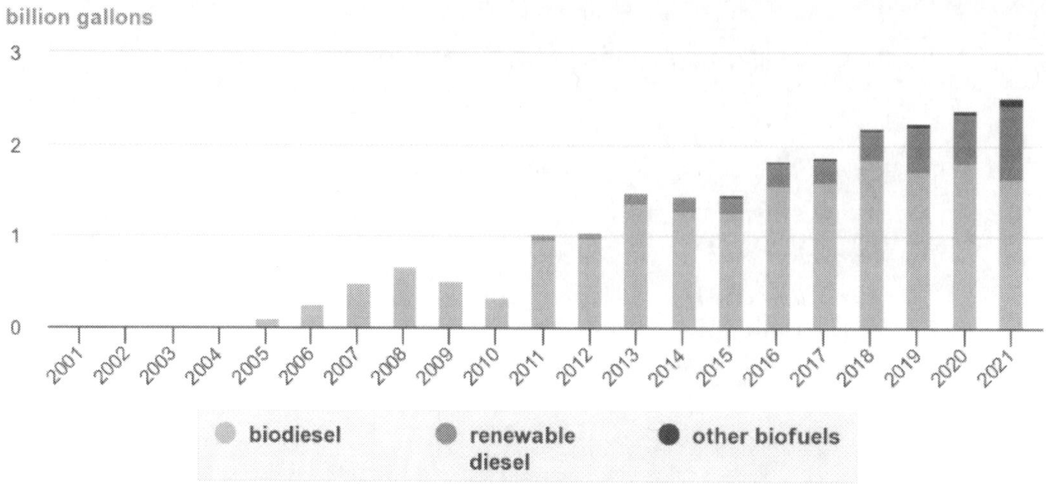

Figure 2. U.S. biodiesel, renewable diesel and other biofuels production from 2001-2021 (source: U.S. Energy Information Administration, *Monthly Energy Review*, March 2022).

Nalco Water is fortunate to work with a broad customer base and our work in additive solutions for biofuels dates back to the earliest days of biodiesel commercialization.[6-7] However, in response to shifting market demands, we have placed particular emphasis in this area over the last 2 years. A family of products has been commercialized that is specifically designed to address the needs of this field of application. We believe these products will enable fuel manufacturers both to reliably process biofeedstocks in their operations and generate quality fuels that are suitable for use in consumer markets.

Our studies have shown that in many cases the preferred chemical additives used for the treatment of biofuels are substantially similar to those used in petroleum-derived fuels. For example, despite the differences in the chemistry of biofuels and petroleum-derived fuels, the underlying mechanisms required to achieve needed anti-static properties or anti-wear (lubricity) properties do not change depending on the source of the hydrocarbon. Dosage requirements may be different, and those differences will be discussed, but the fundamental role of the chemistry remains the same.

There are other cases, however, where the differences in the chemical properties of the feedstocks and fuels have a more significant impact on the preferred chemical treatments. One example of this relates to cold flow properties (e.g., CFPP, cloud point, pour point). In this case, compositional differences in the feedstocks and fuels can have a significant impact on both the properties of the untreated stream as well as the response of the stream to various treatments. We have found that treatments that may be preferred for use in petroleum-derived streams are not necessarily preferred for use in biologically-derived hydrocarbon streams.

Results and Discussion

We began by looking at the fundamental chemical and physical properties of a variety of feedstocks that are commonly used in the production of both biodiesel and RD. As can be seen in Table 1, these properties can vary significantly depending on the feedstock and the processing conditions that were used in its production. In the case of crude tall oil (generated as a byproduct of the Kraft papermaking process), contamination with sulfur, metals, and acidic species have a significant impact on the properties of the produced stream. Conversely, a feedstock like palm oil is considerably less chemically contaminated, but can contain very high levels of insoluble solids (e.g., from agricultural residue) that would be expected to cause fouling in refinery operations. These results indicate that each bio-based feedstock must be characterized and understood independently and cannot be treated as drop-in replacements for one another.

Table 1. Summary of experimentally measured critical properties for various bio-based feedstocks.

Properties	Crude Tall Oil	Soybean Oil	Tallow	Palm Oil
TAN (mg KOH/g)	>130	<0.1	16	7
Sulfur (ppm)	2100	<6	93	<10
Nickel (Ni) (ppm)	20	1	0.126	< 0.5
Vanadium (ppm)	70	0.1	<0.018	<10
ASTM Solids (Lbs / 1000 bbls)	1	<1	<1	140
Bromine number (Cg Br/g)	70	58	36	21

These variations in biofeedstock properties serve as a microcosm of the variability in potential operational challenges for fuels and feedstocks whether sourced from renewable sources or petroleum sources. Table 2 provides an overview of these differences, although specific needs must be determined and addressed on a case-by-case basis. Experience with these various feedstocks and fuels has revealed common trends regarding treatment needs, and we anticipate those trends will continue and be understood more deeply as this market grows.

Table 2. Potential areas of concern for various feedstocks and fuels that are addressable via treatment with chemical additives (✓ = an area of concern, consider treatment; ✗ = not usually an area of concern).

Area of Concern	Feedstocks		Fuels		
	Biofeedstocks	Crude Oil	Biodiesel	Renewable Diesel	Petroleum Diesel
Cold Flow (CP, CFPP, PP)	✓	✓	✓	✓	✓
Biofouling	✓	Rare			
Oxidative Stability	✓	✓	✓	Rare	✓
Conductivity	✗	✗	✓	✓	✓
Lubricity	✗	✗	✗	✓	✓

With these broad and yet disparate needs in mind, our work in this area has focused on identifying chemical treatment solutions that will enable reliable processing of the feedstocks and end use of the fuels. Examples of these solutions will be discussed below.

Focusing on the area of finished fuels, RD is expected to have poor lubricity and conductivity properties owing to its highly paraffinic nature. As shown in Figure 3 below, we have observed this experimentally for representative RD streams. Treatment is needed in order to meet the specifications established by ASTM D975 or EN 590. We have identified lubricity improvers (VIA7300 and VIA7301) and a conductivity improver (VIA7400) that can meet those needs at reasonable dosages. Treatment in this manner enables RD to be used as a drop-in replacement for conventional ULSD derived from petroleum sources.

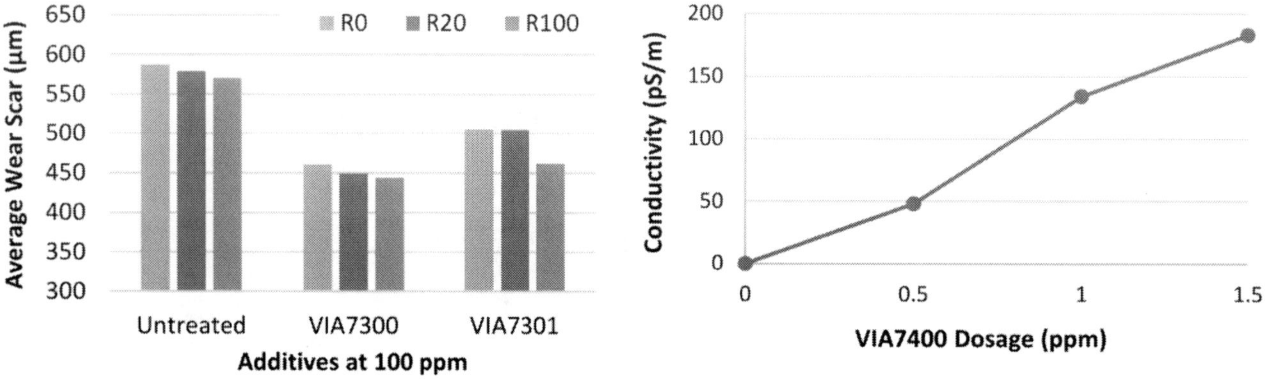

Figure 3. Lubricity properties (left) and conductivity properties (right) for representative treated and untreated renewable diesel streams.

Cold flow properties (i.e., cloud point, cold filter plugging point, and pour point) are another area of concern, but are not a regulated specification to which fuel producers must adhere. Rather, these are often guided by internal guidelines or specifications that depend on the location(s) of a fuel's production and end use. Moreover, cold flow properties vary significantly depending on the feedstock and the processing conditions (e.g., extent of isomerization, hydrocracking). We have identified a variety of additives that can improve cold flow properties of RD blends (Figure 4, left) as well as pure RD (Figure 4, right). With the right selection of product, significant improvement is often possible, thereby enabling a fuel to be used in colder climates.

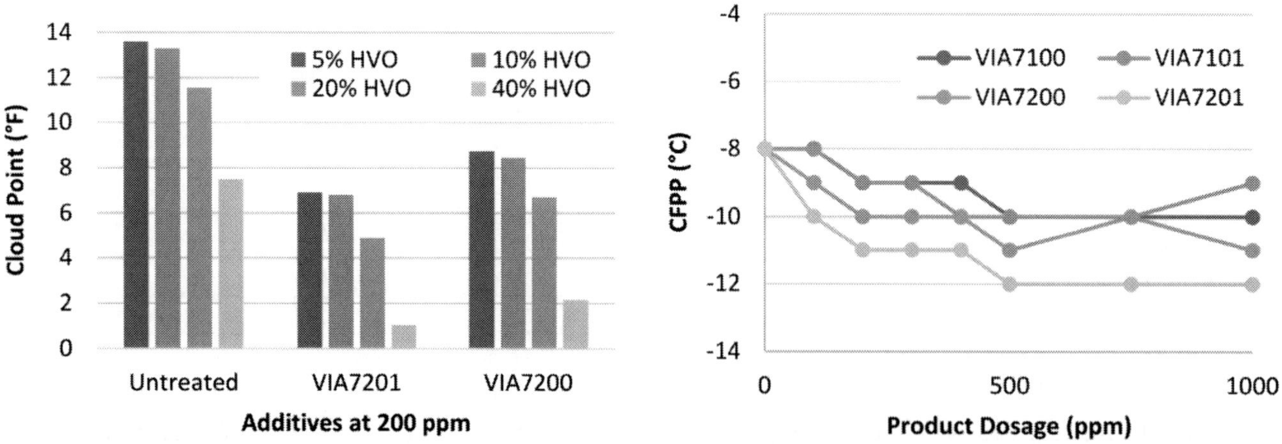

Figure 4. Cold flow properties of renewable diesel blends (left) or pure renewable diesel (right) before and after treatment with either cloud point depressants (left) or CFPP depressants (right).

The results discussed thus far have focused on RD, but as noted previously biodiesel continues to hold a significant portion of the renewable fuels market and is chemically very different from RD or conventional ULSD. We therefore devoted a portion of our efforts to understanding the properties and treatment needs that may be unique to biodiesel.

As shown in Figure 5, biodiesel cold flow properties do not always meet required or desired specifications but can be improved through appropriate treatment with a cold flow improver. Moreover, we have found that the same family of cold flow improvers described previously in the context of RD (VIA7100, VIA7101, VIA7200, and VIA7201) are also applicable for biodiesel. Precise product selection, dosage, and the extent of improvement may be different between RD and biodiesel, but it is useful to have a common set of choices regardless of fuel type.

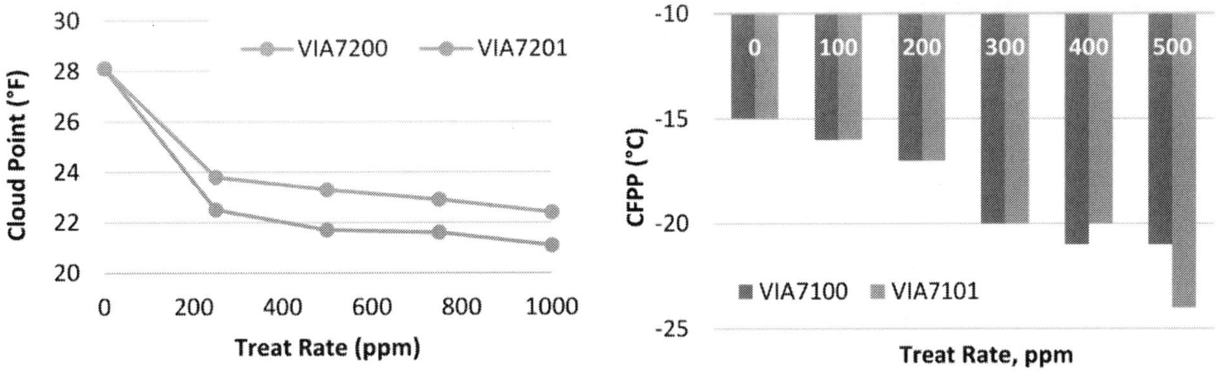

Figure 5. Cold flow properties of pure biodiesel (B100; left) or a biodiesel blend (B20; right) before and after treatment with either cloud point depressants (left) or CFPP depressants (right).

Another potential area of concern for biodiesel that merits particular attention relates to the oxidative stability of the finished fuel. While RD typically exhibits high oxidative stability due to the aggressive hydrotreatment process used in its production, biodiesel does not. Olefinic unsaturation in the feedstock is preserved through the transesterification process, resulting in fuels that are prone to oxidative degradation (e.g., gums formation). The Rancimat test (as described by EN 15751) has been established as a means to quantify the oxidative stability of biodiesel, and the result is expressed as an "oxidative stability index" (OSI) in units of hours. Longer OSIs correspond to increased oxidative stability.

We have found that although untreated biodiesels exhibit relatively low OSI values, oxidative stability can be significantly improved by treatment with an appropriate antioxidant. As shown in Figure 6 below for a representative biodiesel sample (derived from fish oil, which contains a high degree unsaturation), a variety of antioxidants may be considered, each with distinct active ingredients. All of the products improved the stability of the biodiesel, but some performed better than others. Final selection of a product and dosage would likely depend on several factors, including whether a precise OSI specification needs to be met, relative costs of the products, and potential regulatory constraints around the various active ingredients.

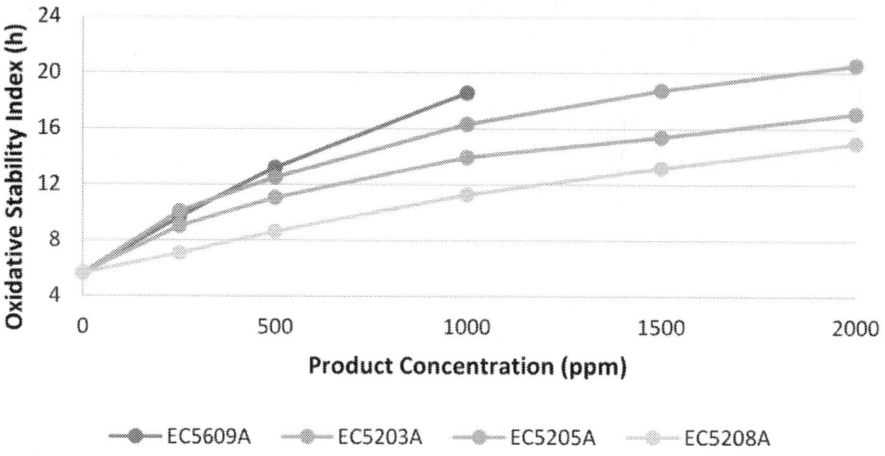

Figure 6. Oxidative stability of a representative biodiesel sample upon treatment with various antioxidants, as measured by Rancimat (EN 15751).

Conclusions

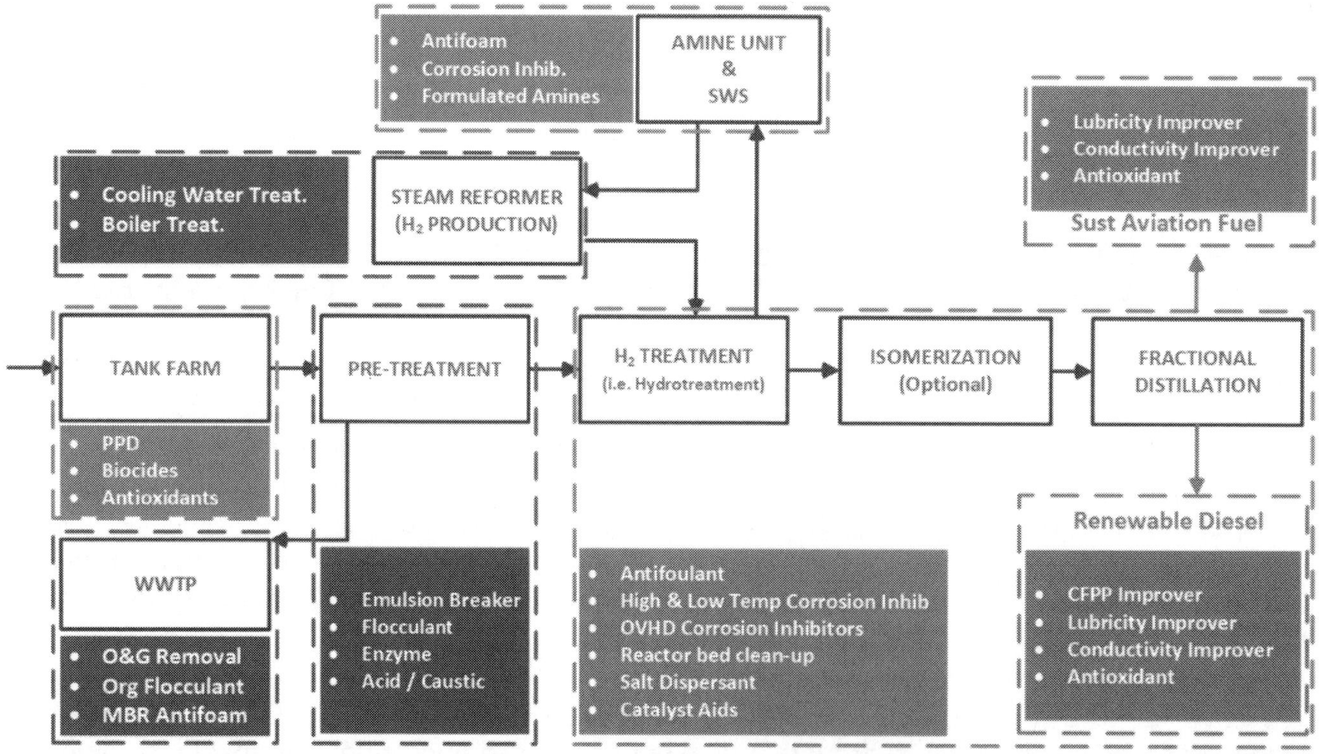

Figure 7. Schematic overview of a renewable diesel (RD) plant and various process treatments and finished fuel treatments that may be needed or desired.

Here we have discussed a variety of challenges and concerns related to the processing and handling of bio-based feedstocks and finished fuels, as well as a suite of chemical treatment solutions that can be used to improve properties that do not meet required specifications. Wide variability in the properties of the feedstocks (depending on the biogenic source) and the fuels (depending on the processing conditions; e.g., RD versus biodiesel) precludes the possibility of a "one size fits all" treatment approach. However, Nalco Water has a broad range of treatment options available, including many that were not discussed herein, such as water treatment chemistries, process antifoulants and corrosion inhibitors, as highlighted in Figure 7 above. Informed investigation of a customer's process and known challenges will enable suitable selection of products to address those needs, while simultaneously avoiding unnecessary treatments and costs.

References

1. A. Sonthalia, N. Kumar "Hydroprocessed vegetable oil as a fuel for transportation sector: a review." *J. Energy Inst.*, 92, **2019**, 1-17.
2. V. Rathore, B. L. Newalkar, R. P. Badoni "Processing of vegetable oil for biofuel production through conventional and non-conventional routes." *Energy Sustain. Dev.*, 31, **2016**, 24-49.
3. B. Boonrod, P. Prapainainar, V. Varabuntoonvit, K. Sudsakorn, C. Prapainainar. "Environmental impact assessment of bio-hydrogenated diesel from hydrogen and co-product of palm oil industry." *Int. J. Hydrogen Energy*, 46, **2021**, 10570-10585.
4. "Biofuels Explained." *U.S. Energy Information Administration*, https://www.eia.gov/energyexplained/biofuels/biodiesel-rd-other-use-supply.php.
5. "EIA projects U.S. renewable diesel supply to surpass biodiesel in AEO2022." *U.S. Energy Information Administration*, https://www.eia.gov/todayinenergy/detail.php?id=51778.
6. K. Peyton, T. McGinnis, P. Bureman "Preventing Sediment Formation in Stored Biodiesel Fuel Blend." *Biodiesel Mag.*, **2008**.
7. T. McGinnis, K. Peyton "Analytical Characterization of Sediment Formed in Soy Oil-Based B20 Biodiesel." *J. ASTM Int.*, 7, **2010**, JAI102584.

635c CFD Modeling Development of Onboard Absorptive Hydrogen Storage on Coal-Derived Activated Carbon for Light-Duty Vehicles

Subhodeep Banerjee, Ki-Joong Kim, Christopher Matranga, Mehrdad Shahnam, William Rogers

Interest in hydrogen as an alternate fuel source has led the automotive industry to search for new power sources for automobiles. The onboard storage of hydrogen represents the current bottleneck in the widespread deployment of hydrogen vehicles and remains a major technological challenge. The U.S. Department of Energy (DOE) has set a technical target of 12 bar and 85 degrees Celsius as the maximum pressure and temperature for onboard storage, alongside goals and time-gated benchmarks for gravimetric and volumetric capacity. Both compressive and cryogenic storage of hydrogen require immense pressures and come with high associated costs and safety concerns. On the other hand, adsorptive storage on activated carbon beds shows promise in addressing the need for safe, reversible hydrogen storage that meets the technical targets set by the DOE.

In this work, computational fluid dynamics (CFD) models of the physisorption of hydrogen on a fixed bed of activated carbon are developed comparing the Eulerian two-fluid model (TFM) approach and the porous media (PM) approach in the commercial CFD solver Ansys Fluent. The different approaches are validated against the bench-scale experiment and PM simulations of Ye et al. [1] and Xiao et al. [2] and contrasted in terms of accuracy and cost and applicability to larger-scale systems. The issue of a creeping error in the calculation of adsorbate mass inside the tank is highlighted and investigated in detail and the causal factors are enumerated. It is shown that despite the reduced numerical complexity of the PM framework, it requires a much smaller time step to accurately account for the mass accumulation and thus incurs a higher simulation cost, and TFM is selected as the optimum simulation methodology.

The validated TFM case is extended to predict the adsorptive storage performance of an activated carbon derived from Blue Gem coal, an exceptionally high-grade coal found in Tennessee and Kentucky with minimal sulfur and ash content. The novel sorbent material, designated BGC-KNa_C850, is developed and characterized by the National Energy Technology Laboratory's Materials and Manufacturing Division. The bench-scale results with the novel sorbent provide a foundation for scale-up and optimization studies for tank design and operating conditions to establish the feasibility of onboard adsorptive hydrogen storage at vehicle scales.

[1] Ye, F. et al. (2012) Implementation for model of adsorptive hydrogen storage using UDF in Fluent. *Physics Procedia*, 24, pp. 793–800.

[2] Xiao, J. et al. (2010) Simulation of heat and mass transfer in activated carbon tank for hydrogen storage. *Int. J. Hydrogen Energy*, 35, pp. 8106–8116.

635f Investigating the Performance of the Pyrolysis Process for Different Feedstocks

Hasan Al-Abedi, Haider Al-Rubaye, Joseph Smith

The significant increase in energy demand, the rapid depletion of conventional energy resources like fossil fuel, and the dangerous environmental effects have boosted the development of renewable and clean energy resources. Biomass conversion is considered one of the promising renewable energy sources in the future since it is abundant and sustainable. Cellulosic (woody) biomass is a renewable feedstock that has the potential to be further converted to energy, high-value chemicals, and biofuels by either gasification or pyrolysis. However, the Biocrude oil produced from biomass has undesired physicochemical properties that prevent it from becoming a useful fuel without further treatments. The main undesired properties are related to the high proportion of reactive oxygenated compounds such as aldehydes, ketones, and acids. One of the promising solutions to mitigate the oxygen level is by using a blend of feedstocks due to the synergy effect.

This research focuses on investigating biocrude oil enhancement technology by utilizing several organic materials as a feedstock for the crude oil production process. The research has four main phases, thermogravimetric and kinetic analysis for individual feedstocks, thermogravimetric and kinetic analysis for a blend and their synergy effect impact, the experimental production of crude oil, and the analysis of the results.

The kinetic analysis of Corn stover, MSW, coal, and oil shale pyrolysis has been investigated after the TGA analysis of that feedstocks was performed. The TGA was done at a certain condition of heating rate, and maximum temperature. This was done as part of quantifying the feedstocks by calculating the water content, volatile content, fixed carbon content, and ash content. Furthermore, the results from the TGA analysis have been used to conduct the kinetic analysis for the feedstock degradation process by using a specified kinetic model. **The one-step global model** and **two-parallel reactions model** are considered the most used for organic feedstock pyrolysis prediction. For MSW, coal, and Oil shale, one step global model was used to predict the kinetic parameters for their pyrolysis reaction. The two-parallel reactions model was applied for corn stover pyrolysis due to the different nature of biomass degradation.

The kinetic analysis results showed that the activation energy for Corn stover, MSW, coal, and oil shale are 18179.11 J/mol, 71294.21 J/mol, 30932.2 J/mol, and 90135.4 J/mol with frequency factor values of 208.1723, 266.5895, 3.594037, 361.5594, respectively.

635g Total Acid Number Reduction of Renewable Thermal Deoxygenation (TDO) Oil for Coprocessing in Refineries

Sampath Karunarathne, Matthew J. Kline, M. Clayton Wheeler

Co-processing biocrude in existing petroleum refineries is an efficient way to produce renewable transportation fuels. It also provides biofuel tax incentives and a potential method for competitive production prices of biofuels. Currently, the major obstacles in the co-processing of biocrude with petroleum crude are high acidity, high oxygen content, and miscibility of biocrude with petroleum crude. The acidity of crude, measured by Total Acid Number (TAN) is an indication of the potential for corrosion problems in a refinery. The presence of oxygen compounds in crude oil has a deleterious effect on the lifetime of hydrotreating catalysts. Typical refinery prefers crude oil with less than 0.5mg KOH/g of TAN and free of oxygen compounds.

In this work, different chemical and catalytic processes are investigated to evaluate the potential of crude Thermal DeOxygenation (TDO) oil for co-processing in a petroleum refinery. TDO oil is a type of synthetic crude produced by a non-catalytic thermochemical conversion of neutralized biomass acid hydrolysate. TDO oil is fully miscible with petroleum products and typically has a low TAN (1.9 mg KOH/g) and low oxygen content (< 5wt%) compared to pyrolysis oil. The presence of naphthols, phenols, and ketones contributes to the acidity and oxygen content of the biocrude.

TDO oils used in these experiments are produced using a mixture of levulinate and formate salts in a pilot-scale continuous flow reactor at 450 °C and atmospheric pressure. Hydrodeoxygenation, neutralization, and ionic liquid extraction methods are used to reduce the TAN below 0.5mg KOH/g and remove oxygen compounds in the crude TDO oil. Hydrodeoxygenation reactions are carried out using catalysts (e.g., Ni/SiO_2-Al_2O_3) in a bench-scale continuous flow reactor at different reaction temperatures and weight hourly space velocities. Neutralization of crude TDO oil is performed with a dilute caustic wash using alkali metals (e.g., Na^+, K^+, Ca^{2+}) at different concentrations. Ionic liquid (IL) extractions are carried out using Imidazolium-based ILs at different ILs/oil ratios. TDO oils are characterized by GC-MS and FTIR to determine the differences in chemical composition. TAN and oxygen content of oils are measured according to standard ASTM methods (ASTM D664 and ASTM D5291 respectively).

65a Carbon-Free Platinum Bismuth Alloy Nanoplatelets with Enhanced Activity and Platinum Stability in Operating Proton Exchange Membrane Fuel Cells and Electrolyzers

Junchuan Fang, Pranjul Mani Dubey, Anastasios Angelopoulos

Hydrogen-powered proton exchange membrane fuel cells (PEMFCs) provide a clean power source for electric vehicles, domestic power plants, and other applications. Their high power density, high energy-conversion efficiency, low working temperature, and low environmental impact can significantly reduce reliance on fossil fuels. However, long-term electrode durability under operating conditions remains a significant challenge to fuel cell commercialization. Corrosion of the underlying electrocatalyst carbon support is a key issue at potentials greater than 0.9 V. Precious metal Pt dissolution and aggregation dominates at lower voltages. Both phenomena yield reduced catalyst area and loss of performance. Alloying Pt with a nonprecious transitional metal (M = Fe, Co, Ni, Cu, and Zn etc.) has been shown to enhance Pt electrocatalytic activity and stability in lab-scale tests. In the past decade, numerous studies have shown that platinum-based binary alloy electro-catalysts, such as PtFe, PtCo, PtNi, PtZn, and PtCu, exhibit enhanced oxygen reduction reaction (ORR) activity in acid electrolyte at beginning of life. However, the use of carbon supports during synthesis of these materials is ubiquitous and loss of activity under operating condition remains unresolved.

In this work, we report on the ORR activity of carbon-free Pt2Bi alloy nanoplatelets (PtBi NP) synthesized via a sustainable, all-aqueous colloidal synthesis route which are found to exhibit previously unrealized long-term stability toward Pt area loss. Beginning of life specific and mass activities of 1.81 A m^{-2}Pt and 402 A g^{-1}Pt were achieved at 0.9 V in perchloric acid; which are 7.1 and 3.1 times higher than those of commercial Vulcan-supported Pt/C, respectively. More significantly, after 10000 potential cycles in accelerated durability testing, the PtBi NP demonstrate no degradation in mass activity, half-wave potential, Tafel slope and electrochemical surface area (ECSA). Such behavior is attributed to a shift in the Pt d-band center observed via XANES which weakens oxygen binding to the catalyst surface and reduces Pt corrosion, consistent with theoretical DFT calculations. The oxophobic PtBi NP also exhibit favorable oxygen evolution reaction (OER) activity for electrolyzer operation with small charge transfer resistance, over-potential, and Tafel slope. Operating performance and durability data consistent with fundamental laboratory findings is presented at 25 cm2 scale.

65b Proton-Conducting Solid Oxide Electrolysis Cells Based on Doped-$BaZrO_3$

Clarita Regalado Vera, Hanping Ding, Jagoda M. Urban-Klaehn, Meng Zhou, Hongmei Luo, Dong Ding

Hydrogen production using proton-conducting solid oxide electrolysis cells (p-SOECs) offers promising advantages due to the reduced operating temperatures of the cells. The efficiency of the p-SOECs depends greatly on the proton-conducting solid electrolyte. However, proton conductors can suffer from electronic conduction and reduced proton conduction at the p-SOEC conditions resulting in low electrochemical performances and low Faradaic efficiencies. In this work, doped-$BaZrO_3$ is studied as an electrolyte for p-SOECs, including a new approach to studying the defect chemistry of proton conductors. The effect of doping on the hydration of the material is revealed, and the improvement in the proton conduction is shown for new doped-$BaZrO_3$ compositions. Finally, the electrolysis performance and Faradaic efficiencies are also studied at reduced temperatures proving the feasibility of new doped-$BaZrO_3$ compositions.

65c Design and Development of a Kilowatt Scale Stack for Carbon Dioxide and Carbon Monoxide Electroreduction

Sean Overa, Feng Jiao

The electrochemical conversion of carbon dioxide (CO_2) to value-added products is a developing approach to offset CO_2 emissions. Previous research has driven ambient temperature electroreduction of CO_2 to single carbon products, such as carbon monoxide (CO) and formate, to technology readiness levels > 4. Higher value multi-carbon (C_{2+}) products, such as ethylene and acetic acid, still suffer from low durability, low sustainability, and high energy demand when produced via direct CO_2 electroreduction. Therefore, an alternative to direct CO2 reduction is to undergo tandem CO_2-to-CO and CO-to-C_{2+} electroreduction. Tandem systems have been shown to have higher selectivity towards single products, specifically acetate, and improved conversion of CO_2-to-C_{2+} products. Our previous efforts focused on CO electroreduction have demonstrated that specially designed CO electrolyzers can continuously produce >25 wt% acetate and a maximum of 60 wt% acetate stream at >70% molar selectivities. Additionally, we coupled this CO electrolyzer with an ambient CO_2 reduction electrolyzer which converted >25% of fed CO_2 to acetate, a significant improvement over previous CO_2 electrolyzers. While performance metrics for these systems have greatly improved, scaling tandem systems to industrial levels still presents a considerable challenge. Here we present our recent work on scaling this tandem CO_2-to-CO-to-C_{2+} system from the commonly investigated small scale (< 10 W total power) to a more industry viable scale (~ 1 kW). We will first focus on our scaling efforts of the CO electrolyzer from the small scale of 5 cm^2 to the medium scale of 25 cm^2, where modification to the cathode and membrane allowed for over 130 hours of continuous operation at a total current of 7.5 A with no degradation in cell performance. We will then highlight our development of a 10 cell stack (1000 cm^2 total) capable of converting CO_2 to CO, CO to C_{2+} products, or tandem CO_2-to-CO-to-C_{2+} at operating powers >0.5 kW. Our investigation into parameters to improve system durability such as cooling, flow pattern design, gas/liquid feed patterns, membrane, and electrodes will also be discussed. Knowledge gained from the development of this stack system will allow for accelerated future development of tandem CO_2 electroreduction at pilot and industrial scales.

65d Modelling the Electrochemical Reduction of CO_2 Under Gas-Liquid Taylor Flow

Isabell Bagemihl, Chaitanya Bhatraju, J Ruud Van Ommen, Volkert van Steijn

Electrochemical reduction of CO_2 using renewable energy sources is a promising avenue for sustainable production of bulk chemicals. However, CO_2 electrolysis in aqueous systems is severely limited by mass transfer leading to performance parameters insufficient for industrial application. A strategy to enhance mass transfer is to actively introduce CO_2 as a gas flow, while keeping diffusion paths between reactants and catalyst short, by separating the gas flow from the liquid electrolyte through a gas diffusion electrode. However, flooding, salt formation, or drying out of the gas diffusion electrode are common problems and present a major obstacle towards commercialization. A promising reactor concept that enhances mass transfer without the complexity of a gas diffusion electrode is a structured reactor operated under gas-liquid Taylor flow. While this is a proven concept to enhance mass transfer for heterogeneous catalysis in flow cells literature on this approach for electrochemical processes is scarce and the effect of process conditions on the reactor performance are poorly understood.

In this work, we propose a tubular cell design, inspired from the field of fuel cells, for CO_2 electrolysis with a zero-gap membrane electrode assembly. We further, develop a simple analytical model revealing how reactor performance in terms of faradaic efficiency and current density is governed by the key features of Taylor flow. These features film thickness, bubble velocity, and volume fraction of CO_2 bubbles over aqueous electrolyte, for a given cathode potential.

We find that the film thickness and the volume ratio of CO_2/electrolyte fed to the reactor significantly affect the limiting current density and the faradaic efficiency. Additionally, we find industrially relevant performance with faradaic efficiencies (> 90 %) at current densities of up to 500 mA cm^{-2}, when operating the reactor at elevated pressure beyond 5 bar. This demonstrates the general potential of this reactor concept to overcome mass transfer limitations in the field of electrolysis. We compare our predictions with numerical simulations, showing good agreement for a large window of operation conditions, illustrating when the simple predictive expressions for the current density and faradaic efficiency can be applied.
We expect that the simple predictive expressions are instrumental in guiding experimental studies and reactor design choices.

65e Modular Dynamic Modeling of Electrochemical Reactors

Dominik Bongartz, Luisa Carola Brée, Katharina Maria Ebeling, Matthias Leitl, Georgia Ioanna Prokopou, Jan Raphael Seidenberg, Alexander Mitsos

Meeting global climate targets requires significant reductions in greenhouse gas emissions in all sectors. For the chemical industry, this entails a shift from fossil raw materials and energy sources to renewables. In this context, Power-to-X solutions that convert, e.g., water and carbon dioxide into valuable products, are gaining increasing attention [1]. Such solutions also enable sustainable solutions in other sectors by producing synthetic fuels, or supporting the integration of renewables in the power sector by providing load flexibility. The heart of almost all Power-to-X solutions are electrochemical reactors that convert electrical energy into the valuable product or an intermediate. As such, the importance of electrochemical reactors is expected to increase significantly in the future, regarding both the number of applications and their scale.

While modeling of electrochemical reactors has received ample of attention in literature [2], models able to describe dynamics are still rather scarce [3]. However, dynamics of electrochemical reactors (as well as the surrounding processes) play a key role in light of the volatility of renewable power sources. Furthermore, dynamic models could facilitate parameter identification and hence aid understanding of the underlying physical effects by separating slow and fast processes [3]. Finally, the increasing number of applications of electrochemical reactors calls for tools to support rapid development of models for different reaction systems and setups.

In this contribution, we present a library for modular, dynamic modeling and simulation of electrochemical reactors in Modelica. This library extends our previous work on modular modeling of CO_2 electrolyzers [4] to provide improved modularity and flexibility in modeling and cover a broader range of applications. Models are provided that describe the chemistry, mass and energy balances of different components such as flow channels or gas diffusion layers, as well as the electrical circuit of such reactors. The models are either spatially lumped or of intermediate fidelity, i.e., rather coarse one- or two-dimensional spatial discretization to maintain computational tractability for dynamic simulation and optimization of entire reactor setups. Beyond the model structure, we present example applications including water electrolysis and electrochemical hydrogen separation and compression.

References

[1] J. Burre, D. Bongartz, L.C. Brée, K. Roh, A. Mitsos, *Chem. Ing. Techn.* 92 (2020), 74.

[2] P. Olivier, C. Bourasseau, P.B. Bouamama, *Renew. Sustain. Energy Rev.* 78 (2017), 280.

[3] B.J.M. Etzold, U. Krewer, S. Thiele, A. Dreizler, E. Klemm, T. Turek, *Chem. Eng. J.* 424 (2021), 130501.

[4] L.C. Brée, M. Wessling, A. Mitsos, *Comput. Chem. Eng.* 139 (2020), 106890.

65f Tailoring Electrode Microstructures for High-Performance Redox Flow Batteries through Non-Solvent Induced Phase Separation

Charles Wan, Rémy Jacquemond, Antoni Forner-Cuenca, Yet-Ming Chiang, Fikile R. Brushett

Porous carbon electrodes play critical roles in electrochemical systems by supporting numerous functions which include providing catalytic sites to facilitate faradaic reactions, distributing liquid reactants and products, and conducting electrons and heat.[1] The surface chemistry and microstructure of the porous carbon electrode is particularly relevant for redox flow batteries (RFBs), wherein dissolved redox active species are forced through porous electrodes within a flow cell, oxidizing and reducing on the electrode surface during operation. RFBs hold promise for long duration energy storage for the electric grid due to their decoupling of energy and power, long service life, and scalability.[2] However, contemporary RFBs are prohibitively expensive, incentivizing further cost reduction for widespread deployment; lowering reactor cost by augmenting power output is a promising approach to bridging the economic gap. While commercial electrodes are functional, they possess suboptimal surface chemistry and microstructure for existent and emerging RFB chemistries.[3] Furthermore, the physical properties (i.e., porosity) of commercial electrodes are spatially homogeneous across the through-plane direction of the electrode, although the optimal porosity profiles balancing fluid dynamics, mass transport, and available surface area may be chemistry-specific and non-uniform.

In this presentation, we will systematically demonstrate non-solvent induced phase separation (NIPS) as a facile and versatile method to engineer interconnected porous microstructures with unique porosity profiles unattainable in commercial electrodes.[4] Combining spectroscopic characterization and fluid dynamic measurements with electrochemical evaluation in two common aqueous redox systems (i.e., $Fe^{2+/3+}$ and all-vanadium), we will illustrate the viability of NIPS as a platform for developing structure-function design principles in RFB electrodes. From these findings, we will highlight pathways towards high-performance electrodes, investigating the role of electrochemically accessible surface area and porosity gradients in the through-plane direction in these custom material sets. Further, we will show that the synthetic method can be modified to generate dense and planar carbon films, enabling opportunities to extract fundamental rate constants on bottom-up designed materials using conventional electroanalytical approaches. While the primary focus of this talk will be on RFBs, the methods and findings described are generalizable to convection-driven electrochemical reactors that require spatially engineered electrode layers.

Acknowledgments

This work was supported as part of the Joint Center for Energy Storage Research, an Energy Innovation Hub funded by the U.S. Department of Energy, Office of Science, Basic Energy Sciences. CTW acknowledges additional funding from the National Science Foundation Graduate Research Fellowship Program under Grant No. 1122374. Any opinions, findings, and conclusions or recommendations expressed in this material are those of the author and do not necessarily reflect the views of the National Science Foundation.

References

1. K. J. Kim, M.-S. Park, Y.-J. Kim, J. H. Kim, S. X. Dou, and M. Skyllas-Kazacos, *J. Mater. Chem. A*, **3**, 16913–16933 (2015).

2. B. Dunn, H. Kamath, and J.-M. Tarascon, *Science*, **334**, 928–935 (2011).

3. A. Forner-Cuenca and F. R. Brushett, *Current Opinion in Electrochemistry*, **18**, 113–122 (2019).

4. C. T.-C. Wan andamp; R. R. Jacquemond, Y.-M. Chiang, K. Nijmeijer, F. R. Brushett, and A. Forner-Cuenca, *Advanced Materials*, **33**, 2006716 (2021).

256a Reservoir Quality Indicators and Relative Thermal Maturity Classification Using Image-Based and Bulk Rock Characterization

Shannon Eichmann, David Jacobi, Poorna Srinivasan, Jennifer Rodriguez

Mature source rocks consist of a complex pore space where nanoscale pores are hosted by a heterogeneous matrix with varying minerals and organic components to generate a natural nanocomposite with chemical and textural heterogeneity. Organic and inorganic nanopores in these rocks both store and transport hydrocarbons. The organic-hosted pores are generated during thermal maturation of the rock as hydrocarbons are generated. This paper presents a method using image-based characterization to provide relative comparisons of reservoir quality between wells and a method to combine image-based and crushed rock analyses to compare source rock maturity. The results and workflow presented impact special core analysis for unconventional reservoirs and reservoir quality assessment and can complement characterization obtained by other methods. Scanning Electron Microscopy (SEM) is used for source rock characterization to understand rock texture and compositional variations, porosity, and pore sizes. However, despite having significant benefits to characterization, obtaining quantitative results by SEM is time consuming and costly, and therefore the number of images collected per well is generally limited. Recent advances in image processing make obtaining quantitative data from images more accessible. This improves our ability to gather more image-based data on multiple wells for integration with larger scale measurements. Carbonate rich source rocks were sampled from several wells for SEM imaging. Prior to imaging, the samples were mounted, polished, and ion milled. Large field-of-view SEM images were collected and segmented using supervised machine learning to label the pores, fractures/cracks, organics, high density minerals, and matrix minerals. Finally, post-processing methods were used to correct mislabeled components. The relative amount of organic-contained porosity to total porosity (R1) and the relative amount of organic content to total porosity (R2) were calculated for each sample. Porosity was also obtained using the Gas Research Institute method. Pyrolysis was used to determine the hydrogen index. In general, samples with organic content of >2 vol% and total porosity >1% are considered to be of higher quality than others. Beyond these typical bounds, the results show that the R1 and R2 ratios from image-based data can be used to define samples that indicate potentially better quality. When compared across several wells of similar maturity, these quality metrics and criteria can be applied to suggest wells of better quality. Finally, by comparing image-based data to that measured at larger scales, thermal maturity indicators can also be provided.

256b Benefits of Removing Nitrogen before the Cold Section in Baseload LNG Plant

Ajinkya Pal, Iftekhar Karimi, Easa Al-Musleh

Natural gas (NG) is the cleanest fossil fuel and is regarded as the bridge for the transition from non-renewable to renewable energy sources. Transport of natural gas over long distances as liquified natural gas (LNG) at -160°C and atmospheric pressure, is the most common and economical option. The NG feed from the field undergoes a series of energy-intensive up-gradation and liquefaction processes to produce the final LNG product, falling in with customer specifications. A realistic baseload LNG plant consists of hot and cold sections, with the former consisting of inlet processing, acid gas removal (AGRU), dehydration, and sulfur removal units. The cold section operates at sub-ambient conditions and consists of the liquefaction unit, NG liquids (NGL) recovery unit, nitrogen removal unit (NRU), helium removal unit (HRU), and consumes more than 60% of the total power requirement.[1]

Nitrogen is an inert gas and does not contribute to the heating value of gas. Moreover, LNG calls for stringent and lower nitrogen specifications, as >1mol% nitrogen content in LNG tanks can cause safety hazards and risk of rollover.[2] The separation of nitrogen from feed NG is implemented by designing separation processes that exploit differences in the molecular, thermodynamic, and transport properties of nitrogen and hydrocarbons (methane). As most of the physical properties of methane and nitrogen are comparable, designing a cost-effective nitrogen removal process is challenging.[2] Nonetheless, the difference between the boiling points (B.P) of methane (B.P ~ 111.7 K) and nitrogen (B.P ~77.3 K) allows their separation by distillation process only under cryogenic conditions.[3] Most LNG plants remove nitrogen near the tail of the plant after the liquefaction process. Hence nitrogen passes through the entire system as dead weight. Therefore, the presence of nitrogen in the NG stream increases the energy requirements and reduces the plant production capacity.

To address the above concerns we propose upfront nitrogen removal (UN_{rem}) as a promising alternative to enhance the feasibility of the LNG process. The concept of UN_{rem} is based on removing nitrogen from the NG stream before it enters the cold section. The UN_{rem} will create extra volume in the cold section, which will be replaced by hydrocarbons. Hence the UN_{rem} would increase the plant capacity while reducing the power required to liquefy NG. We first carry out simulation and optimization of the cold section for different percentages of UN_{rem} to quantify the improvements in the production rate (PR) and specific power consumption (SPC). Then we review all possible physical and chemical-based nitrogen separation technologies to select an efficient UN_{rem} process that can either be retrofitted in an existing LNG plant or can be included in a new LNG plant.[4] Finally we carry out a thorough techno-economic analysis to estimate the cost of retrofitting UN_{rem} technology into an existing LNG plant and evaluate the impact of different percentages of UN_{rem} on the overall cost of LNG production.

We computed the energy consumption, PR, and SPC of seven different UN_{rem} scenarios and compared them with the conventional cold section (base case). The results indicated that 87.5% UN_{rem} increased the LNG PR by 4.4% and decreased the total energy consumption and SPC by 0.241% and 4%, respectively, in comparison with the base case.[5] After reviewing and comparing various physical and chemical-based nitrogen separation technologies based on energy consumption, feasibility, and scalability, cryogenic distillation, membranes, and lithium cycle (Li-Cy) were shortlisted as potential UN_{rem} technologies. Though distillation and membranes have high technology readiness level (TRL), their energy consumptions are high. Hence Li-Cy was selected as UN_{rem} technology for the techno-economic analysis. Results show that 87.5% UN_{rem} would decrease the cold section CAPEX and OPEX by 3.2 % and 2.3% respectively compared to the base case. The CAPEX and OPEX of the lithium cycle increase linearly with the UN_{rem} percentage. The total annual cost (TAC) and LNG energy content increased by 12.3% and 4.5% respectively over the base case for 87.5% UN . This leads to a 7.4% COE increment over the base case rem

for 87.5% UN_{rem}. The profit calculation for LNG price of 15$/MBTU showed a 4.27% or 101.74 million $/year increase in profit for 87.5% UN_{rem} as compared to the base case.

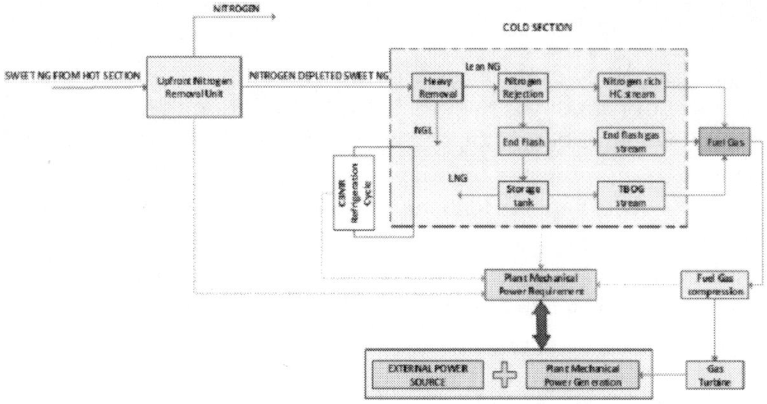

256c Modeling Uncertainty and Impact of Non-Darcy Flow Regimes on Unconventional Well Performance.

Harun Ates, Shannon Eichmann, Kanhaiyalal Patel, Ravi Vaidya, Rabah Mesdour

Tight nano-pores in unconventional reservoirs can cause deviations from Darcy's law where permeability becomes a strong function of pore size and pore pressure. Many studies have been conducted to quantify and correct for nano-pore effects on laboratory permeability measurements to account for changing flow regimes from slip flow to Knudsen diffusion. Several studies have noted that corrections to Darcy flow effects may be also important when analyzing well production behavior. However, there are no studies or guidance available to suggest under what reservoir or well conditions the non-Darcy flow regimes become important for a well's performance. This paper illustrates how one can easily model the impact of non-Darcy flow regimes using a commercial simulator.

As pore dimensions gets smaller, the interaction of gas molecules with the pore walls becomes increasingly dominant as compared to interactions between gas molecules. This increases gas mobility; hence, lab permeability measurements may not be representative of actual reservoir conditions where; slip flow and Knudsen diffusion is important. This workflow uses the predictions from generalized Non-Darcy flow correction models as a function of Knudsen number as input into a flow simulator. As a result, one can model impact of changing flow regimes from Darcy flow to Knudsen diffusion. A stochastic approach is then used to assess the impact of non-Darcy flow on unconventional wells through flow simulations accounting for uncertainties in pore sizes, reservoir properties and well constraints.

This workflow provides guidance and risk assessment for the significance of gas slippage and Knudsen diffusion in production behavior and recovery of unconventional wells. The developed workflow was tested on both synthetic and real field datasets where the results showed that Non-Darcy flow regimes may have significant impact under certain reservoir conditions. This is a novel stochastic approach for the integration of data from pore level to reservoir scale to assess the impact of nano-pore effects on unconventional well performance. Importantly, it does not require that the simulator code be modified and can be performed using existing commercial simulator functionalities.

256d Elucidation of Non-Catalytic Ethylene Polymerization Reactions through Computational Study

Alexander Shaw, Grant Marsden, Linda Broadbelt

Ethylene is a highly important industrial precursor chemical that is used primarily in the synthesis of various polymers. Ethylene, whether obtained directly from natural gas sources or synthesized from other hydrocarbons, is transported primarily through pipelines under high pressure conditions.[1] In order to mitigate ethylene polymerization within pipelines during transport, it is essential to understand the mechanisms by which the polymerization occurs. While ample research effort has been applied to understanding the catalytic formation of various oligomers from ethylene,[2,3] both using homogeneous and heterogeneous catalysts, knowledge of how such oligomerization may occur in the absence of catalysts is still lacking.

Owing to the intense focus on catalytic ethylene polymerization, coupled with a scarcity in recent quantum chemical investigations into the topic, the mechanisms of ethylene polymerization are still not well understood. In this work we employ density functional theory calculations to elucidate the pathways by which oligomers are formed and to better understand the temperature and pressure dependence of the product distributions. We begin by modeling and comparing several bimolecular reactions of ethylene, identifying those which may serve as initiators for subsequent chain growth reactions. From these initiation reactions we identify two key routes for the generation of biradical species, one through homolytic cleavage of cyclobutane, and the other through cleavage of cyclobutene. Lastly, we explore how, through intramolecular rearrangements, intermolecular hydrogen abstractions, and free-radical polymerization reactions, these biradical species facilitate both the formation of oligomers and the generation of new radical species through more facile routes than available during the initiation period. The results of this work can help guide transport conditions in order to minimize oligomer formation during ethylene transport in pipelines.

All calculations were conducted in the gas-phase using the M062X meta-hybrid functional with the Def2-TZVP basis set of Alrichs and co-workers. The UltraFine integration grid and default optimization convergence criteria were used throughout. Dispersion was included in the form of Grimme's D3 empirical dispersion correction without any damping. Contributions to the enthalpy of the system from frequency modes below 100 cm-1 were adjusted using the quasi-harmonic (QH) correction of Head-Gordon. Contributions to the entropy from the same small vibrational modes were adjusted using the QH method of Grimme. Both methods were used as provided in the GoodVibes software. GoodVibes was also used to scale all vibrational frequencies by a factor of 0.971, as recommended by the work of Donald Truhlar at the University of Minnesota.

References:

1 J. M. Bielicki, R. S. Middleton, J. S. Levine and P. Stauffer, *Energy Procedia*, 2014, **63**, 7215–7224.

2 J. Malinowski, D. Jacewicz, B. Gawdzik and J. Drzeżdżon, *Sci. Rep.*, 2020, **10**, 2–7.

3 R. Y. Brogaard and U. Olsbye, *ACS Catal.*, 2016, **6**, 1205–1214.

256e Functionalized Iron Oxide (Fe_3O_4) Nanoparticles As a New Type of Recyclable Hydrate Dispersant

Yuanxing Zhang, Ning Wu, Amadeu Sum

The agglomeration of gas hydrates in flowlines may disrupt the normal operation and even block them, causing enormous financial loss. The current solution is to inject low-dosage hydrate inhibitors or surfactant-like anti-agglomerant. The concerns about using such chemicals are mainly about the cost, toxicity, and environmental impact. More recently, nanoparticles have been investigated to be a potential candidate for dispersing small hydrate particles and allowing them to flow in a slurry form. In this work, we coated 20 nm iron oxide particles with different hydrophobicity and studied their impact on hydrate dispersion. The response of iron oxide nanoparticles to the magnetic field was also utilized for recycling. Both on a microscopic scale with in-situ observations at ambient conditions and on a macroscopic scale with high-pressure rock-flow cell experiments, we explored the performance of nanoparticles with different extent of surface modifications. With proper surface modifications, the iron oxide nanoparticles prevented small hydrate particles from agglomeration. Furthermore, we investigated efficient particle recycling strategies with a magnetic field. The nanoparticles were fully detached from the oil-water interface by adding a tiny amount of surfactant (0.5% to the water phase). After drying, they can be readily reused, reducing the cost and minimizing environmental concerns.

319a Hydrogen Generation from Natural Gas Using Autothermal Chemical Looping Systems: Process Simulation and Heat Integration of Two Process Configurations

Fanhe Kong, Yitao Zhang, Pinak Mohapatra, Andrew Tong, Liang-Shih Fan

Chemical looping is an advanced material and energy conversion technology that can achieve both high level process intensification and efficiency. To analyze chemical looping processes, it is essential to include process conditions that are realistic and comparable to those that are expected of in industrial systems. Relevant variations in these conditions as occurred in bench versus industrial scale systems include isothermal versus adiabatic operation of the reactors, and local versus global process heat integration. Naturally, the types of reactors employed dictate how the reactor operation is to be conducted from the heat integration viewpoint in the overall process arrangement. As an example, in industrial applications, a fluidized bed reactor is operated near uniform temperature conditions. A fixed bed or a moving bed reactor on the other hand is typically operated adiabatically, and thus under the autothermal operation, the non-isothermal condition prevails, leading to different strategies for process simulations and heat integration requirements. This study presents the chemical looping process simulation based on a moving bed reactor used as a reducer for two H2 generation process configurations under autothermal operating conditions. The two process configurations are represented by the two-reactor (reducer – combustor followed by the water-gas shift reaction) and the three-reactor (reducer – oxidizer – combustor with water splitting for H2 generation in the oxidizer) chemical looping systems with each configuration producing H2 in a different operating scheme. The simulation results are compared with the conventional steam-methane reforming (SMR) system as a baseline case to underscore the attractiveness of the chemical looping configurations. Specifically, for each configuration, the parametric study under the adiabatic conditions is used to optimize the operating conditions that can satisfy the heat balance requirements and can achieve a maximum H2 yield. The exergy analysis indicates that the two-reactor chemical looping and three-reactor chemical looping systems can achieve, respectively, a 4.0% and 11.4% increase in relative percentage in the overall process exergy efficiency over the conventional steam-methane reforming system.

319b Catalytic Conversion of Methane to Hydrogen and Carbon Nanotubes By Microwave Irradiation

Changle Jiang, Brandon Robinson, Sonit Balyan, Jianli Hu

Carbon nanotubes supported Ni-Pd or Ni-Cu bimetallic catalysts were adopted to dehydrogenate methane and produce carbon nanotubes and hydrogen. The catalytic reactions were driven by microwave irradiation only. Both catalysts showed methane conversions over 35% at 550 °C. Ni-Cu displayed a better dehydrogenation performance and a lower cost when compared to Ni-Pd. The TEM and Raman characterizations of spent catalysts have indicated that the solid products were carbon nanotubes. The activation energy of 10Ni-1Cu-CNT was calculated as 24.8 kJ/mol under microwave irradiation, which was much lower than that of conventional heating (45.5 kJ/mol). The results have indicated that 10Ni-1Cu-CNT catalyst could be further scaled up and promoted for simultaneously converting methane to carbon nanotubes and hydrogen under microwave irradiation. Additionally, carbon nanotubes could be sold to reduce the overall cost for hydrogen production. Technoeconomic analysis indicates that the technology has the potential to lead to \$ 1/kg H_2.

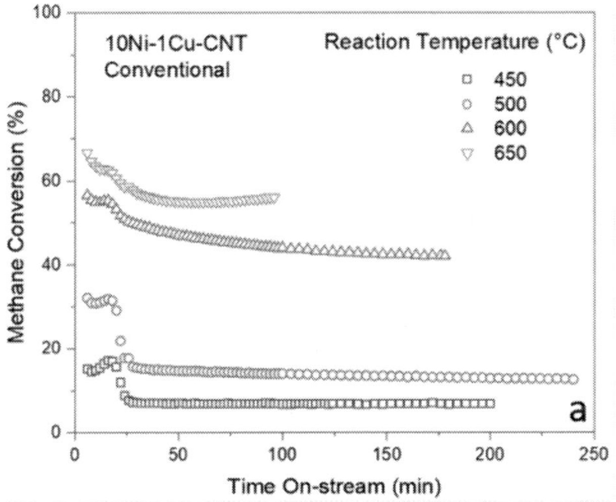

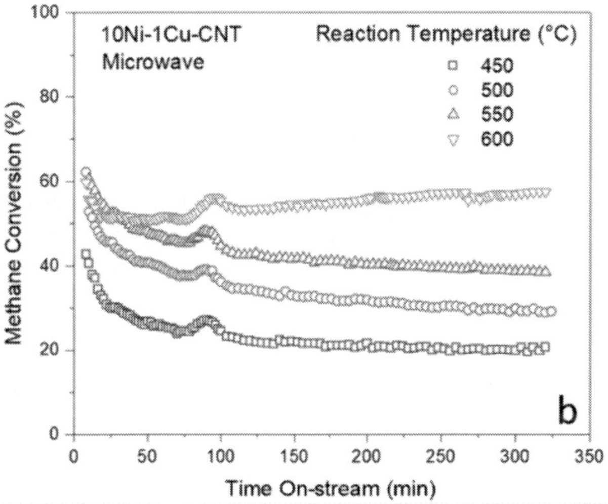

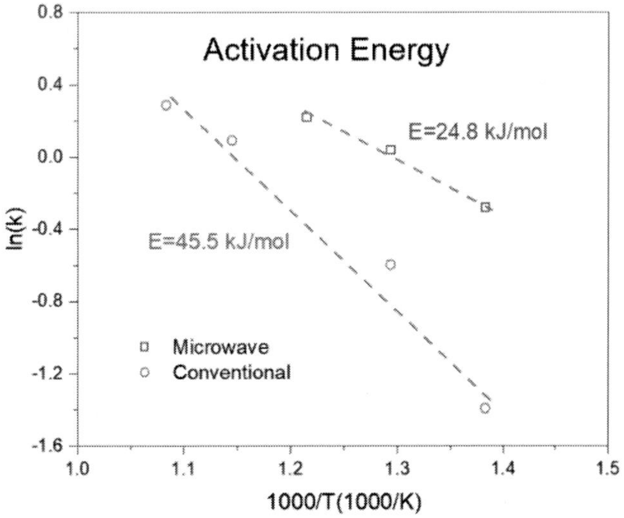

319c Exploring Production Aspects of Blue Hydrogen and Valuable CNTs from Natural Gas in a Fluidized Bed Reactor

Kaushal Parmar, Kamal Pant, Shantanu Roy

The need for clean, sustainable energy sources is increasing due to environmental fouling from fossil fuel emissions. Hydrogen is one of the best clean fuels with high energy density and no adverse environmental effects. The most common hydrogen production method, i.e., Steam Methane Reforming (SMR), emits a high amount of CO_2 (1300 times more than the product). Hydrogen produced in this way is called grey hydrogen. Exploring blue hydrogen and green hydrogen is essential considering current environmental concerns. Thermo-Catalytic Decomposition of methane (TCD) is an attractive approach for producing blue hydrogen. The energy requirement of the TCD process is quite similar to commercial SMR (37.8 kJ/mol H_2 vs. 41.3 kJ/mol H_2). Furthermore, the TCD process avoids energy-intensive units, i.e., pressure swing adsorption for CO_2 separation, because it produces CO_x-free H_2 and solid carbon. Different forms of carbon (i.e., CNTs, graphene) can be made by optimizing the shape and size of the metal crystals of the catalyst.

In this context, methane decomposition in a fluidized bed reactor was investigated. A transition metal-based alumina-supported catalysts were prepared and tested for methane decomposition in the fluidized bed reactor. These catalysts obtained high methane conversion (>80%) and multi-walled bamboo-shaped carbon nanotubes. Prepared catalysts and CNTs were characterized by TPR, BET, FE-SEM, HR-TEM, Raman spectroscopy, TGA, etc. Hydrodynamic aspects of the fluidized bed reactor were also taken into consideration.

References:

1. K. R. Parmar, K. K. Pant, S.Roy, Energy Convers Manag 2021, 232, 113893.
2. S. K. Saraswat, K. K. Pant, Int. J. Hydrogen Energy 2011, 36, 13352.

319d Compact Flex-Fuel Processors for Syngas and Hydrogen Production

Hani Hawa, Saurabh A. Vilekar, Christian Junaedi, Subir Roychoudhury

Precision Combustion, Inc. (PCI) has developed a novel reactor and process for low carbon intensity blue-hydrogen (H_2) production from natural gas (NG) capable of operating with high efficiency (HHV basis). The design is based on PCI's short contact time Microlith® catalyst. Both oxygen- and air-blown systems with integrated steam generation were examined. Key goals were to achieve high energy efficiency with CO2 capture. A process-intensified, simple reactor that minimizes external heat input and with internal steam generation was designed, fabricated, and tested. A core innovation was PCI's catalyst technology for coating nano-powder catalysts on unique Microlith short contact time, high surface area substrates, enabling several advantages notably small size, rapid transient response, high selectivity and efficiency, and robust operation. An extremely compact reforming catalyst substrate reforms NG into a mixture of H_2 and CO. The catalyst and reformer technology, in combination with process intensification of heat exchangers, permits extremely compact system footprint. This leads to significant reduction in CAPEX and OPEX compared to traditional steam methane reforming (SMR) with post combustion carbon capture. Higher system efficiency of the proposed process lends to reduction in the amount of CO_2 produced per unit of H_2 generated to further reduce carbon capture costs. We will report data from sub-scale and pilot testing and compare results with equilibrium predictions as well as metrics from state of the art commercial H_2 generation via steam methane reforming.

319e Ni-MgO Based NOVEL Catalysts Prepared By Combustion Synthesisfor CO_2 and Methane Reforming

Vardan Danghyan, Alexander Mukasyan, Anand Kumar, Eduardo E. Wolf

In this talk we present mechanistic aspects of the solution combustion synthesis (SCS) method to prepare of high surface area Ni-MgO solid solutions as active and stable catalysts for the dry reforming of methane (DRM) with CO_2, two major greenhouse gases.

In SCS, an aqueous solution containing a metal nitrate precursor (oxidant) is mixed with glycine (fuel) and dried to obtain a thick reactive gel, which is preheated until at a certain temperature the highly exothermic combustion reaction is initiated resulting in the formation of metal oxides. A modification of SCS was used by impregnating the reactive solutions into a cellulose paper, referred as paper assisted combustion synthesis (PACS).

During the synthesis, the combustion reaction exothermicity, temperature, and front propagation velocity were monitored by high-speed Infrared thermography and differential scanning calorimetry (DSC). The Ni-MgO formed solid solutions were characterized with textural (BET, SEM), structural (XRD, TPR, TEM), and surface analysis (XPS) techniques. Activity and stability of catalysts during the DRM were studied as a function of reaction temperature, time on stream (TOS), carbon deposition, and the reaction turnover frequency (TOF) was evaluated at differential methane and CO_2 conversions.

NiOMgO catalysts with 10, 20 and 30% Ni concentration were first prepared by SCS. which formed NiOMgO solid solutions with surface areas of up to 37 m^2/g, which were inactive for DRM. Upon reduction at 600 ^oC, Ni atoms segregated to the surface of the solid solutions, forming crystallites containing the active sites. On 20%Ni-MgO reduced catalyst, methane conversion reached about 50% at 600 ^oC, however, there was also carbon accumulation of about 80 wt.% after about 2 hours of TOS.

Unlike to SCS, PACS synthesized Ni-MgO solid solutions exhibited much higher surface areas, partly because during combustion Mg^{2+} ions strongly catalyzed cellulose pyrolysis, creating a porous media, which acted as a template to the NiO-MgO solid solutions. The 10%NiOMgO-PACS catalyst exhibited the highest surface area of 153 m^2/g compared to only 23 m^2/g prepared by SCS method. At 600 ^oC, the 10%NiOMgO-$PACS_R$ catalyst exhibited methane conversion close to equilibrium and significantly inhibited carbon deposition (~3 wt.% during 24 h TOS), thus increasing the catalyst's stability. The higher activity and stability of the 10%Ni-MgO catalyst is attributed to the higher dispersion and smaller size of Ni crystallites formed during the PACS synthesis, and the formation of a higher amount of Ni^{3+} surface site defects (2).

1) V. Danghyan, A. Kumar, A. Mukasyan and E. E. Wolf[1], Appl. Cat. B 2020, 119-156.

2) V Danghyan, T. Orlova, S. Roslyakov, E. Wolf, A. Mukasyan. Combust. Flame 2020, 221, 462-475.

319f Solution Combustion Synthesis of Iron-Based Alumina Nanocomposites for Microwave-Assisted Thermocatalytic Dehydrogenation of Fossil Fuels

Zachary Chanoi, Victoria Reyes, Evgeny Shafirovich

Iron-based alumina nanocomposites are promising materials for microwave-assisted thermocatalytic dehydrogenation of fossil fuels, which has been proposed for the production of hydrogen with no CO_2 emissions. These powders both absorb microwave radiation and catalyze the dehydrogenation. Solution combustion synthesis (SCS) is an attractive technique for the fabrication of these materials because it has been used for synthesis of many nanoscale oxides. However, SCS can be carried out using different organic fuels and different heating equipment, and it is important to determine which fuel and heating mode generate materials with the best properties for dehydrogenation, such as maximized specific surface area.

In the present work, the iron-based alumina nanocomposites were obtained by SCS using two different fuels (citric acid and glycine) and two different heating modes (a hotplate and a muffle furnace). The precursors/oxidizers were iron nitrate and aluminum nitrate. They were mixed with the fuel according to stoichiometry and dissolved in water. Heating the solution resulted in combustion synthesis. The products were characterized by X-ray diffraction analysis, Brunauer-Emmett-Teller surface area analysis, laser diffraction particle size analysis, scanning electron microscopy (SEM), and energy dispersive X-ray spectroscopy (EDS).

XRD analysis revealed phases of hercynite, with no significant effect of the heating mode. The fuel change slightly influences the XRD pattern: products obtained using citric acid have less-crystalline spectra. SEM-EDS of the products revealed flaky structures with increased concentrations of iron in several spots on the particle surface. The products obtained using citric acid as the fuel have a narrow particle size distribution as compared to those produced using glycine, independent of the heating mode. The specific surface area of iron-based alumina nanocomposites obtained by SCS was dramatically increased by using citric acid instead of glycine, and additionally by using a muffle furnace instead of a hotplate.

This material is based upon work supported by the Department of Energy under Award Number DE-FE0032086.

319h Exploring the Potential of Ni_3S_2 Based Sulfur Looping Process for Efficient Production of H_2 from H_2s: A Thermodynamic and Kinetic Study

Anuj Joshi, Kalyani Jangam, Zain Mohammad, Liang-Shih Fan

Hydrogen Sulfide (H_2S) is a poisonous gas that is released during processing of fossil fuels. Despite the rise in utilization of renewable-based energy sources, non-renewable hydrocarbon fuels dominate the world's energy production scenario and are slated to do so until 2050. Thus, a high amount of H_2S generation will be continued for next couple of decades, making its treatment critical. Claus process is state-of-the-art H_2S abatement technology; however, it suffers from various issues such as unfavorable reaction thermodynamics, high capital cost and requirement of an air separation unit (ASU). Moreover, the Claus process is ineffective in extracting value from H_2S, as it converts the 'Hydrogen' content of H_2S into low-value steam (H_2O) instead of valuable hydrogen (H_2) gas. To counter this, a sulfur looping process scheme utilizing Ni_3S_2 as a sulfur carrier was established for effectively decomposing H_2S into H_2 and S. It involves two sub-steps – sulfidation and regeneration. In the sulfidation step, Ni_3S_2 reacts with H_2S forming NiS and H_2, and in the subsequent regeneration step, sulfur is recovered as NiS is thermally decomposed back into Ni_3S_2 in presence of CO_2. The sulfur looping scheme has been demonstrated at a laboratory scale fixed bed reactor, wherein ~80% H_2S conversion was achieved at a gas hourly space velocity of 5000 hr^{-1} [1].

In this work, process analysis of sulfur looping was performed, and the results were compared against Claus process to establish the commercial feasibility of the process. Parametric studies were carried out on the sulfur looping process in ASPEN to obtain the reaction and feed conditions for both the sulfidation and regeneration steps. The thermodynamic results obtained were validated using relevant thermogravimetric experiments and X-ray diffraction analysis. After the conditions were determined through the combination of process simulations and experiments, system evaluation parameters such as overall H_2S conversion, H_2 and sulfur yield were calculated for the sulfur looping process and compared against the baseline Claus process. The thermodynamic performance was further quantified by carrying out energy and exergy analysis. Results indicate that for >99% H_2S conversion, ~2.4% higher sulfur yield is obtained for the sulfur looping process with 100% selectivity towards H_2 generation (no steam formation). Moreover, a significant increase of ~19% and ~8% in overall energy and exergy efficiencies, respectively was observed for sulfur looping process due to H_2 production and requirement of fewer processing units. This work further corroborates the superior potential of the novel sulfur looping scheme to replace the existing Claus process as it overcomes its drawbacks and creates a valuable product from a toxic industrial waste. The findings from this study can be used to develop an economical and environmentally sustainable process.

[1] Jangam, Kalyani V., Anuj S. Joshi, Yu-Yen Chen, Shailaja Mahalingam, Ashin A. Sunny, and Liang-Shih Fan. 'Synergistic decomposition of H_2S into H_2 by Ni_3S_2 over ZrO_2 support via a sulfur looping scheme with CO_2 enabled carrier regeneration.' *Chemical Engineering Journal* 426 (2021): 131815.

319i Integrated SOLID OXIDE Electrolytic Cell for a NOVEL Hydrogen Production Using Water-Energy Nexus Framework

Lateef Jolaoso, Javad Asadi, Pejman Kazempoor

ABSTRACT

Water and energy are two inseparable resources needed for the existence, wellbeing and economic sustainability by human. The production of energy requires substantial amount of freshwater. There is an increase in energy demand due to several factors like the population growth. During the continuous expansion of an energy sector, the mix of the deployed technologies to produce fuels and generation of electricity controls the associated impact on freshwater resources in that area. This work explores the nexus model by focusing on the design, performance and economic estimation of a novel, highly efficient and environmentally benign hydrogen production process which can be integrated with several power production units, including coal, natural gas-fired power plants and renewable energy sources. The system employs flue gas from coal power plants and various wastewater sources to generate steam as the process's feedstock for the SOEC unit. The SOEC and the balance of plant components are designed, simulated, and integrated using Aspen HYSYS Software. The SOEC model is experimentally and numerically validated, and a piece-wise performance evaluation of the system is carried out. The result of this innovative system shows overall thermal-to-hydrogen efficiency of 56.3% and 61.97% if the heat of the high-temperature exiting species are put to use and the SOEC shows a voltage efficiency of 92.9% which can be improved depending on the energy source and by full energy Integration. The techno-economic analysis of the system gives a Levelized hydrogen cost of $2.99/kg, which can be upgraded to meet the U.S. Department of Energy (DOE) targets. The study offers insights into water-energy nexus for the researchers and policymakers for decision making and technology development.

Keywords: Solid oxide electrolytic cell (SOEC), Hydrogen production, Flue gas, Water-energy nexus, techno-economic analysis (TEA)

INTRODUCTION

The production of hydrogen from renewable and sustainable sources is a key to the green hydrogen economy. Water electrolysis has gained attention and looks promising as the process is greenhouse gas-free and yield 100% pure hydrogen. Even though water is abundant, its accessibility is a great challenge, and its use for this process will undermine water security for domestic and agricultural use. One of the major crises faced in this generation due to its increasing population is the availability of clean drinking water, particularly in developing countries. Even the developed ones can only access 2% of the available water[1]. China, most especially its coal-rich region, is faced with water scarcity[2] and projected to be severe in 2050 and the arid and semi-regions like the Middle East and the western U.S. [3].

There are three major commercialized and conceptualized water electrolytic cells for the generation of hydrogen in recent times [4] which include alkaline electrolytic cell (AEC), the most matured technology [5], polymer electrolyte membrane (PEM) electrolytic cell (PEMEC) and solid oxide electrolytic cell (SOEC). The AEC challenges range from low current density to low operational pressure, which aimed to be solved with the advent of PEMEC[6]. Likewise, the PEMEC suffers from high capital cost, pure water requirement, and increased system complexity, stemming from its operational pressure [6, 7]. However, the SOEC is an evolving and promising technology that is anticipated to overcome the setbacks plaguing these two electrolyzers [8, 9]. Solid ion-conducting ceramics are used as electrolytes in the SOECs, which operate at high temperatures ranging from 650-1000 °C [9]. Several works have been done on this technology both in the laboratories [8] and industries to commercialize it [9]. Kazempoor and Braun[10] modeled and evaluated a SOEC system to serve as a primary system-level performance prediction tool. Several contemporary studies on SOEC focused on hydrogen production directly from water and steam [11-13]. For better system optimization high round-trip efficiency, SOEC has been designed and studied alongside its reverse counterpart, solid oxide fuel cell (SOFC), to form regenerative solid oxide cell (rSOC) [14-16]. To eliminate the uncertainty concerning future cost and high capital cost of this technology, recent literature evaluates its techno-economic viability to encourage investments[17-19].

This work is distinctive in its water electrolysis in that it explores the flue gas from power plant and wastewater as a feedstock into the hydrogen production unit integrated with a different energy source. The net energy and techno-economic analyses are carried out on the entire system for performance evaluation and sensitivity analysis on the system's critical parameters.

SYSTEM AND MODEL DESCRIPTION

The proposed system is shown in fig. 1. It comprises two sections, namely the upper steam generating section and the high-temperature SOEC unit- hydrogen production unit.

In the upper section, high pressurized steam is generated from the wastewater reservoir after it is pumped to the water treatment and recovery unit (WTRU) where it is evaporated at atmospheric pressure and 101 °C. This is mixed with the steam, which is concurrently generated from the power plant flue gas. The flue gas is cooled to 25°C in the condenser for moisture recovery, and the heat regained is used in the WTRU. The recovered water is immediately evaporated and combined with the steam from the WTRU. The mixed stream is sent to the pre-heater to elevate its temperature to 750-850°C before being fed into the SOEC to lower the required entropic heat for decomposition. Air from the ambient environment is compressed, preheated, and supplied to the system to flush out the produced oxygen from the electrolysis reactions in the SOEC. The exiting hydrogen product is stored in a compressed hydrogen tank.

The water decomposition reaction taking place in the SOEC is represented by the anodic oxidation and cathodic reduction reaction below

Planar model configuration was used in the SOEC design with the system comprising of the balance of plant (BoP) such as condenser, heater, compressor, mixer, evaporator, separator, pump, heat exchanger to finally actualize a working model which are thermodynamically model using hysys. All other model assumptions are similar to that of Kazempoor and Braun[10].

RESULTS AND DISCUSSION

This section summarizes some of the important results of this work. The SOEC is in the developing stage, and this novel work is evaluated and validated. The system generated 242400 kg/day of hydrogen from the flue gas from a power plant with a capacity 333.7 MW. The voltage efficiency of the SOEC system is 92.9% at 850 °C, and the overall thermal efficiency of the designed plant is 61.97%, which seems to be very prosperous compared to other plants of the same size. Even though it is shown from hysys that is potential to save over 25% of the utilized energy in the system which reduces cost of production and optimize productivity. To live up to the progressive development in different sources of energy, the model is designed to be integrated with several energy sources. The Levelized cost of hydrogen (LCOH) production with some of these energy sources is evaluated and has shown in fig. 2. The LCOH ranges from \$2.99-5.96/ kg hydrogen with the solar photovoltaic (P.V.) given the cheapest cost assuming the U.S. Department of Energy (DOE) target of 3 cent/kWh by 2020 and beyond. As such, the life cycle analysis (LCA) from cradle to gate for the plant is performed in SimaPro using the solar P.V. and the coal power plant to measure the severity of the environmental impact. Obviously, the COx emission with the coal power plant is higher. However, SimaPro revealed the higher effects of some other critical parameter factors like the human health, ecosystem and particulate matter formation associated with it. This

CONCLUSIONS

The novel model designed in this work shows the possibilities of different energy integrations for hydrogen production and redefined a new approach to water-energy nexus. The LCOH of hydrogen can be as low as \$2.99 per kg hydrogen, and this is a close route to meeting the DOE target for the cost of hydrogen for distribution and as potential energy storage. Further work needs to be embarked on to establish the exact environmental impact of large-scale hydrogen production to meet up the global requirement for sustainability and a green economy.

Acknowledgement

This material is based upon work supported by the U.S. Department of Energy, under Award Number DE-FE0032005.

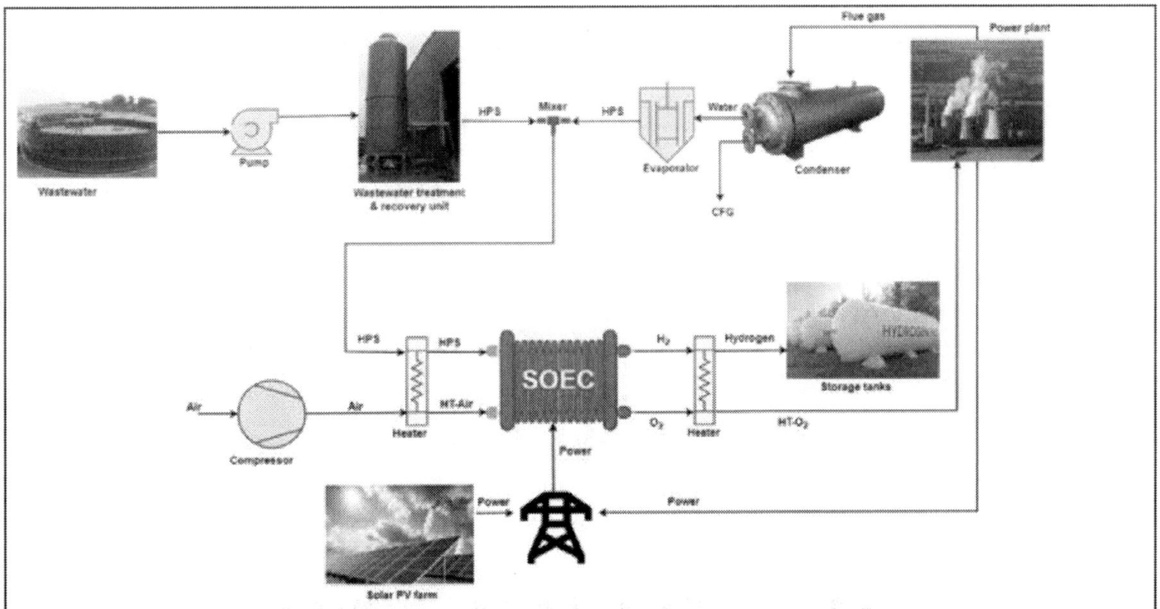

Fig 1: Water-energy integrated system for hydrogen production

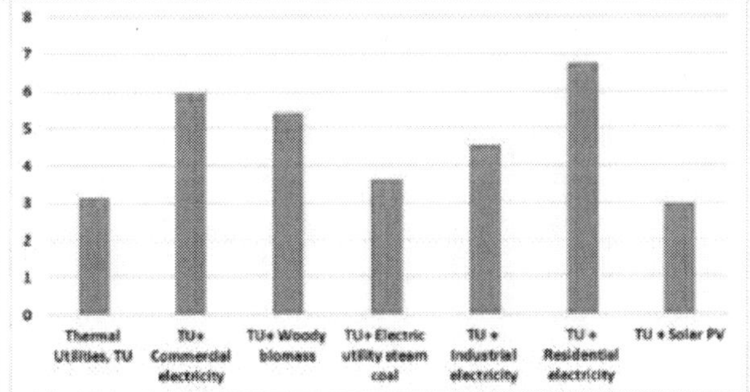

Fig. 2: Levelized cost of hydrogen (in $) with different feedstocks

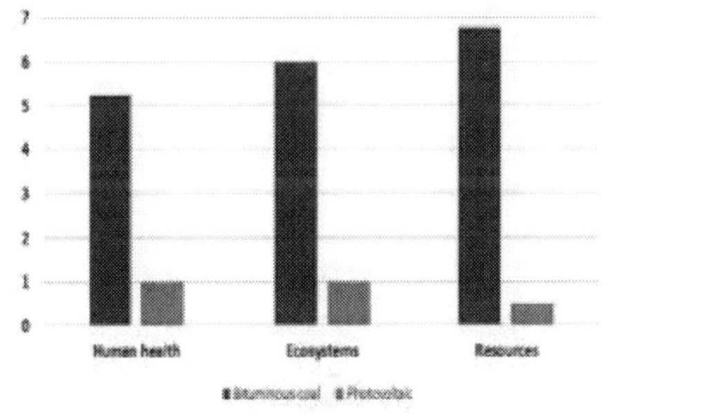

Fig 3: Damage assessment for electrolysis reaction by solar PV and coal power sources

535a Acetylene Powered Thermal Power Plant Boiler

Debshilpi Patra

Coal is depleting at an alarming level due to an increase in demand and increasing load on the generation sector. The present study focuses on the incorporation of alternative fuels along with existing components without making major modifications. Acetylene is a type of alternative and synthetic fuel that is sustainable, energy-conserving, engine efficiency, and needs minor modifications from current boilers. It is easy to produce from simple and rapid hydrolysis reactions of calcium carbide and water at a low cost. As calcium carbide is produced from lime and coke, whose availability is abundant, acetylene gas has the potential of becoming the most prominent source as an alternative fuel for the engine. As Acetylene is a highly combustible gas with a calorific value of 48225 kJ/kg, it can be used in the boiler as an alternative fuel to attain super-critical temperature. At this critical point, the pressure and temperature attained to be 22.1 Mpa and 374°C, respectively.

535b A Novel Fluid System to Integrate Stimulation, EOR, and CO_2 Storage in Shale Oil Reservoirs

Yang Zhao, Yanxu Ding

Due to the ultra-low permeability nature of shale oil reservoirs, the oil production rate is much lower than from conventional hydrocarbon resources. CO_2 has been preliminarily proven an effective way to extract the tremendous unconventional resources. Also, CO_2-EOR is an effective way for CO_2 utilization and storage, thus, helping reduce carbon emission and realize the "carbon neutral" target. A new fluid system was proposed to integrate stimulation, EOR, and CO_2 storage in shale oil reservoirs.

The new working fluid system was established by adding cosolvent and surfactant into supercritical carbon dioxide (SC-CO_2). The fluid was used for stimulation to create hydraulic fracture networks and for soaking to displace oil present in the matrices. Bottle tests were carried out to investigate screen proper cosolvent and surfactant candidates and the stability of the fluid system at supercritical conditions. Phase behavior experiments were conducted to study the minimum miscible pressure (MMP) and the miscibility behaviors with and without cosolvent. Imbibition tests were performed to look into the oil production rate, equilibrium oil recovery and the wettability index.

Results show that the new fluid system was stable under supercritical condition without precipitation. Phase behavior experiment results demonstrate that the addition of cosolvent reduced the MMP and help to achieve miscibility with shale oil. The influence of cosolvent type and concentration was studied. Imbibition tests results suggest that the new fluid system promoted the spontaneous imbibition process. The imbibition reached the equilibrium status in a shorter time. That is, the oil present in the shale matrices was displaced out more rapidly. In addition, a higher equilibrium oil recovery efficiency was obtained with the proposed fluid system. Wettability alteration was observed after exposure to the working fluid. The wettability index was estimated and the results indicate that the shale medium became less oil-wet. The effects of cosolvent type and concentration, and surfactant type and concentration were tested. The system was optimized according to the tests. As a result, the cosolvent facilitated miscibility of CO_2 and shale oil, and the surfactant promoted imbibition of the working fluid into the matrices. The enhanced miscibility and imbibition help CO_2 penetrate and be trapped in the shale matrices (i.e., enhanced CO_2 sequestration).

Systematic experimental and theoretical studies were carried out to test the feasibility of a new working fluid of SC-CO_2/cosolvent/surfactant combination for shale oil development. The work demonstrates the potential of the proposed fluid system in improving the performance of stimulation, economic recovery factor, as well as CO_2 sequestration in shale oil plays.

Key words: Shale oil; Well stimulation; Supercritical CO_2; Fracturing fluid; CO_2 storage

535c Fluoro- Vs Hydrocarbon Surfactants: Synthesis and Application in Enhanced Oil Recovery

Syed Hussain, Muhammad Shahzad Kamal, Ahmad Mahboob, Mobeen Murtaza, Shirish Patil, Emad W. Al Shalabi

Hydrocarbon-based surfactants show high surface tension and high critical micelle concentration (cmc). Moreover, hydrocarbon-based surfactants decompose at high temperatures in oilfield reservoirs and also face stability issues when encountering monovalent and divalent reservoir ions. The harsh environment of oilfield reservoirs causes surfactant degradation and minimizes their ability in lowering surface/interfacial tension and changing the wettability of reservoir rocks.

To address the stability issues related to hydrocarbon-based surfactants, fluorosurfactants were synthesized by a one-step reaction, which was confirmed by NMR, FTIR, and ESI-MS analysis. The heat stability and water solubility tests were performed by Thermal Gravimetric Analysis (TGA) and aging methods. The interfacial tension was measured using the spinning drop method. The cmc and surface tension at cmc were calculated by a force tensiometer (Biolin Scientific). The rheological parameters were identified with the aid of a discovery hybrid rheometer (DHR-3).

The in-house synthesized fluorosurfactants were found to be soluble in water and TGA results exhibited a degradation temperature above 100 ^{o}C which is superior to real oilfield temperature ($\geq$90°C). In addition, the aging technique revealed no precipitation by visual observation after putting the surfactant samples in the oven at 90 ^{o}C for 30 days and it was further confirmed by chemical structure analysis of fluorosurfactant using NMR and FTIR techniques. The surface/interface properties of the fluorosurfactants were comparable to or higher than industrially applied surfactants. Rheological results showed that the storage modulus was decreased by enhancing the surfactant concentration at a lower rate of shear and frequency due to polymer interaction and charge screening.

The in-house synthesized fluro surfactants displayed excellent thermal stability and water solubility and established huge potential in severe conditions of oilfield reservoirs.

535d Fast Pyrolysis of Food Waste in N_2 and CO_2 after Torrefacton in a Bubbling Fluidized Bed Reactor

Seung-Soo Kim, Hoang Vu Ly, Jinsoo Kim, Hyun Tae Hwang

In this work, the mixed food waste (MFW) was pretreated via torrefaction at moderate temperatures (250-275 °C) under an inert atmosphere before fast pyrolysis. The pyrolysis of torrified MFW (T-MFW) was performed in a bubbling fluidized-bed reactor (FDB) to study the influence of torrefaction on the pyrolysis product distribution and bio-oil compositions. The highest liquid yield of 39.54 wt% was observed at a pyrolysis temperature of 450 °C. The torrefaction has a significant effect on the pyrolysis process of MFW. After torrefaction, the higher heating values (HHVs) of the pyrolysis bio-oils (POs) ranged from 31.51 to 34.34 MJ/kg, which were higher than those of bio-oils from raw MFW (27.69-31.58 MJ/kg). The POs mainly contained aliphatic hydrocarbons (alkenes and ketones), phenolic, and N-containing derivatives. The pyrolysis of T-MFW was also carried out under the CO_2 atmosphere. The application of CO_2 as a carrier gas resulted in a decrease in the liquid yield and an increase in the gas product yield. In addition, the carbon and nitrogen content of POs increased, whereas the oxygen was reduced via the release of moisture and CO. Using CO_2 in pyrolysis inhibited the generation of nitriles derivatives in POs, which are harmful to the environment. These results indicated that the application of CO_2 to thermal treatment of T-MFW could be feasible in energy production as well as environmental pollution control.

535e Thermogravimetric and Kinetic Study of *Hemp* and *Milkweed*

Seung-Soo Kim, Jinsoo Kim, Foster Agblevor

Hemp and *Milkweed* are a kind of biomass which can be easily cultivated with rapid growth rate. As a renewable resource, herbaceous biomass feedstock such as *Hemp* and *Milkweed* are promising for the production of bio-oil. Pyrolysis of *Milkweed* was carried out in a thermogravimetric analyzer to understand their pyrolysis characteristics, and pyrolysis kinetics of those materials was studied using a micro tubing reactor. *Hemp* and *Milkweed* decomposed mainly for 250~410 °C and 200~500 °Cat heating rates of 5-20 °C/min, respectively, and those of average apparent activation energy were 100.27 kJ mol^{-1} and 156.43.7 kJ mol^{-1} within pyrolysis conversion of 10~80%. The kinetic parameters of *Hemp* and *Milkweed* pyrolysis were determined using nonlinear least-squares regression of the experimental data assuming first-order kinetics. The kinetic rate constants indicated that the predominant reaction pathway was **A** (Biomass) to **B** (Liquid), rather than **A** (Biomass) to **C** (Gas; C_1-C_4) and/or **B** (Liquid) to **C** (Gas; C_1-C_4) at temperatures of 330-370 °C.

535f pH and Salt-Responsive Dynamic Binary Complexes Based on Supramolecular Complexation with Applications in Unconventional Reservoirs

Bhargavi Bhat, Shuhao Liu, Yu-Ting Lin, Joseph Kwon, Mustafa Akbulut

Hydraulic fracturing of reservoirs has seen tremendous growth in recent years, to keep up with the ever-increasing energy demand. The rheological properties of the fracturing fluid used during the process plays a crucial role in determining the result. A key research problem has been being able to design fluids that carry proppants with enough driving force into fissures while keeping costs reasonable. Hence, several efforts have been active towards development of more efficient and improved fracturing fluids while keeping sustainability in check.

Herein, we present a dynamic binary complex (DBC) solution, composed of a facile self-assembly of a supramolecular assembly of oleic acid and diethylenetriamine under alkaline conditions. With lower than 2 wt% constituents dispersed as an aqueous suspension, a gel-like substance that exhibits high viscosities even under application of shear was formed. Important attributes include dual responsiveness of the rheological properties to salinity and pH, as well as a zero-shear viscosity that can be tuned by a factor of ~280. Also, the gel-like nature was more obvious in the presence of salt which was confirmed by observing the morphology under a microscope. Sand settling tests revealed its potential to hold up sand particles (proppant) for extended periods of time.

In conclusion, this DBC suspension system with a simple assembly process has the potential to be utilized as a smart pH-switchable and salt-responsive hydraulic fracturing fluid. Also, the fact that oleic acid is a natural substance derived from olive oil may add to the sustainability and cost-efficacy of this system.

535g Upgrading Tire-Derived Pyrolysis Oil Via Distillation for Blending with Fuel Oils

Ryan Dantu, Johannes Knoetze, Cara Schwarz

Introduction

Worldwide waste tires are stockpiled resulting in environmental problems. In addition, significant calorific value is captured in these spent tires. A possible solution to the problem of spent tires is to pyrolyze the rubber fraction of spent tires to produce a crude oil commonly known as tire derived oil (TDO), which has a significant calorific value. The production of TDO therefore not only addresses the problem of spent tires but also harnesses the calorific value captured therein and can be used as a fuel source.

During the vulcanization process of the production of tires significant quantities of sulphur are introduced into the rubber to improve the durability of the tires. The sulphur propagates through to the TDO during the pyrolysis process, therefore resulting in a high sulphur content of the TDO (typically 0.82 mass %, as in the sample used here). Further, the TDO has a broad molecular mass range, typically broader than that of crude oil, therefore requiring fractionation before it can be used as a fuel source. The aim of this paper is to present results in the upgrading of TDO via distillation to investigate whether distillation is a viable process to obtain fractions that can be blended with current fuel fractions.

In this work commercial tire derived oil was upgraded via fractional distillation in a batch distillation column. Two sets of experiments were conducted, namely bench scale batch distillation tests at atmospheric pressure and pilot plant scale tests varying the pressure, reflux ratio and heating input, and the results evaluated for application as fuel fractions.

Methodology

The bench scale setup consisted of a single stage 500 mL pot setup containing 100 mL of TDO. The setup operated at atmospheric pressure and five fractions were collected namely distillates at temperatures below 140°C (naphtha), between 140°C and 200°C (Diesel 1), between 200°C and 245°C (Diesel 2) and between 245°C and 300°C (Marine Bunker Oil, MBO), and the bottoms product. The products were analyzed for sulphur content and fuel properties. Five repeats of the experiment were conducted.

The column consisted of a 50 L pot loaded with 15 kg of TDO, a 1.7 m high packed bed (68 mm diameter, 1/4" Rashig rings random packing) and reflux was controlled via a solenoid valve between 5 and 20. The column was operated up to 300 °C and at pressures from atmospheric down to 30 kPa absolute. Pressure was controlled using a 1.5 kW vacuum pump. Experiments were conducted varying pressure, reflux ratio and heating rate (1.24 to 2.48 kW). A central composite design was used to obtain the optimum conditions resulting in 16 experimental runs. Distillate fractions were obtained at temperatures below 140°C (Naphtha), between 140°C and 245°C (Diesel), and between 245°C and 300°C (MBO) resulting in four final products (3 distillate fractions and remaining bottoms). These fractions were analyzed for sulphur content and fuel properties to evaluate their suitability to be blended as fuel fractions.

Results and Discussion

Mass balance results showing the various fractions obtained for the repeats of the bench scale study are shown in Table 1. The results show that significant quantities of each of the fractions can be obtained with the MBO fraction being the larges.

Analysis of the various fractions show that the sulphur content increased with increased boiling point ranges with that of the naphtha fraction being 0.56 wt.% increasing to 0.77 wt.% and 0.72 wt.% in the two diesel fractions and 0.75 wt.% in the MBO. Except for the MBO, these values are well above specification. Further, legislation for MBO specifications is changing and the sulphur content will be in excess of specification.

The other fuel properties (viscosity, density, water content and calorific value) of all fractions are either within specification or close to. Where these properties are not within specification it is believed that finetuning of the process will result in on-spec products.

The results for the masses obtained for each of the fractions for the pilot plant experiments are shown in Table 2. The mass balances in Table 2 typically account for between 92.6 and 96.7 % of the mass. Differences in the mas balance are typical for pilot plant scale studies and can be accounted for through accuracy of measurements, losses through the vacuum systems and column hold-up, amongst others.

Analysis of the naphtha fractions showed that the sulphur content was lower than for the batch distillation runs (0.24 – 0.41 wt.%) but still not within specification. Also, the fuel properties were again either close to or within specification with fine-tuning expected to yield on spec properties. Similar results were obtained for the Diesel fraction with sulphur content between 0.41 and 0.61 wt.% and physical properties close to or within specification.

For the MBO fraction the sulphur content was higher than in batch processes (0.84 to 1.07 wt.%) which is also within current specifications. Further the other properties were also close to or within specification.

Considering the study conducted here the optimum conditions depend on the target product. The most naphtha was obtained at 70 kPa, a reflux ratio of 10 and a 2kW heating rate for 15 L of oil (equivalent to 1.3 approximately $kg.s^{-1}.m^{-2}$). The most diesel was obtained at 101 kPa, a reflux ratio of 10 and a 2kW heating rate (approximately 1.6 $kg.s^{-1}.m^{-2}$) and the most MBO at 101 kPa, a reflux ratio of 20 and a heating rate of 2.48 kW (approximately 1.9 $kg.s^{-1}.m^{-2}$). Further alternative optimums may result if the sulphur content and/or the fuel properties are targeted.

It should also be noted that the experimental results presented here were conducted at a range of pressures but the temperatures at which the cuts were obtained remained the same. Alternative optimums may be obtained if the temperatures ranges are adjusted at the various pressures to align with the vapor pressures.

Conclusions

From this study it can be concluded that TDO can be considered a viable source for fuel fractions. The crude oil can be fractionated into fractions that have similar specifications than various fuel fractions. However, due to the sulphur present due to the vulcanization process, the sulphur content is currently too high. Options for the reduction of sulphur include the pre-treatment of the tire rubber before pyrolysis and the post-pyrolysis desulphurization of the oil.

A significant quantity of further work still needs to be done before viable fractions are produced. In addition to the aforementioned desulphurization, further optimization of the distillation process is also required. Additionally, fractional condensation of the pyrolysis outlet gasses to produce a pre-fractionated TDO should be considered. All these factors need to be considered and optimized for the process as a whole.

Table 1: Experimental results of the bench scale distillation runs showing the wt.% of each of the fractions collected

Run	Naphtha (wt.%)	Diesel 1 (wt.%)	Diesel 2 (wt.%)	MBO (wt.%)	Bottoms (wt.%)
Temperature range	< 140 °C	140 °C – 200°C	200°C – 240°C	245°C – 300°C	> 300°C
1	16	12	12	29	31
2	19	14	15	30	22
3	15	13	15	33	24
4	15	12	14	36	23
5	18	13	14	29	23
Variation	17 ± 1.62	13 ± 1.09	15 ± 1.71	32 ± 3.04	25 ± 3.73

Table 2: Experimental results for the pilot plant runs study showing the wt.% of each of the fractions collected at each of the conditions. Parameters were varied around a center point (0) between high (+1) and (-1) in the form of a center composite design.

Run	Pressure (Pa)	Reflux Ratio (-)	Heating rate (kW)	Naphtha (wt.%)	Diesel (wt.%)	MBO (wt.%)	Bottoms (wt.%)
1	101 (+1)	5 (-1)	1.24 (-1)	1.82	17.65	15.68	61.60
2	101 (+1)	5 (-1)	2.48 (+1)	1.52	18.21	15.35	60.95
3	101 (+1)	20 (+1)	1.24 (-1)	4.21	19.65	24.14	46.56
4	101 (+1)	20 (+1)	2.48 (+1)	5.01	18.56	36.46	35.20
5	30 (-1)	5 (-1)	1.24 (-1)	10.63	20.98	28.98	33.01
6	30 (-1)	5 (-1)	2.48 (+1)	12.65	18.65	28.65	35.10
7	30 (-1)	20 (+1)	1.24 (-1)	15.96	21.89	35.01	20.80
8	30 (-1)	20 (+1)	2.48 (+1)	17.56	23.89	33.15	21.56
9 C	70 (0)	10 (0)	2 (0)	15.98	24.41	30.41	24.60
10	101 (+1)	10 (0)	2 (0)	11.56	18.63	26.32	36.12
11	30 (-1)	10 (0)	2 (0)	17.89	24.51	29.45	21.15
12	70 (0)	5 (-1)	2 (0)	20.58	19.62	24.15	28.60
13	70 (0)	20 (+1)	2 (0)	18.56	21.56	25.48	30.56
14	70 (0)	10 (0)	1.24 (-1)	20.65	19.54	23.48	30.42
15	70 (0)	10 (0)	2.48 (+1)	19.04	20.53	25.14	28.62
16 C	70 (0)	10 (0)	2 (0)	15.65	25.14	28.65	24.60

535h Development of Bioethanol from Molasses: A Modern Approach

Ayush Gondane, Tanashree Paraye, Dr. Utkarsh Maheshwari

Because it is a waste product of the sugar manufacturing process, molasses may be purchased from that business. Molasses do not have a direct use and are thus regarded as a waste product of the sugar manufacturing process. Nowadays, in order to make the process more cost-effective, the generated molasses might be repurposed for another purpose. Bio-ethanol is the preferable product that may be made from molasses. Because of the high sugar content of these products, this method is favored.

To convert molasses into bioethanol, the process known as Simultaneous Saccharification and Fermentation (SSF) is used. The procedure only uses 10-20% of the substrate's total volume.

Biological processes are used in the hydrolysis of sugarcane molasses, with the employment of Aspergillus vinegar and Trichoderma varied, and the fermentation process is carried out utilizing the consortium of bacteria. Saccharomyces cerevisiae-Pichia stipites are a kind of yeast. . At 72 hours fermentation, the bioethanol concentration produced by the SSF techniques was the highest, at 14.38 percent v/v for a 20 percent substrate and a 10 percent consortia.

535i Crude Assays Cut Properties

Andrew Sloley

Crude Assays:

Cut Properties

Andrew W. Sloley

DGI

Bellingham, WA, USA

Crude assays include more than distillation curves. Specific boiling ranges of material in the crude have different compositions. These compositions affect product properties and processing required. A useful crude assay provides information on individual boiling ranges, cuts, separated from the crude by distillation. Properties can be measured on each of these cuts. This analysis looks at consistent trends in these properties relevant to the Crude Distillation Unit (CDU) and Vacuum Distillation Unit (VDU). The CDU and VDU separate components based on boiling point, but more than boiling point is important in them. Both operating costs and selected (non-distillation) product properties are important in CDU and VDU operation.

Operating costs in the crude unit are strongly affected by:

- asphaltenes and other composition related properties in fouling,
- chlorides leading to potential corrosion,
- acid content and corrosion, and
- sulfur compounds and corrosion.

Many products have specifications that are strongly affected by crude unit performance. While no general correlation is available that relates boiling point and the property, general trends are reliable and performance is strongly related to boiling point for a specific crude. These properties may include:

- freeze point for kerosene going to jet fuel,
- luminometer number for kerosene going to jet fuel,
- smoke point for kerosene going to jet fuel,
- cloud point for diesel,
- volatile metals contents for gas oils,
- sulfur, nitrogen, and oxygen for various middle distillate and gas oil fractions,
- viscosity and viscosity index for gas oils (lubricants),
- carbon residue for gas oils and residues, and
- penetration for asphalts (residues).

This paper reviews common trends in these properties in the cuts from crude oil and how actual unit performance, property trends, and property characteristics can affect product qualities and operating costs.

535j Screening Criteria and Experimental Investigation for Enhanced Oil Recovery at Umm Niqa Oil Field-Kuwait

Ahmed AlShammri, Osamah Alomair, Adel Elsharkawy

Oil and gas resources are the primary income sources for several countries, particularly in the Middle East. Moreover, oil consumption is rapidly growing owing to the increased global demand for energy, which encourages oil-producing countries to optimize their production. In general, most conventional oil resources have been extracted, leaving vast non-conventional crudes trapped in reservoirs, the extraction of which is challenging and requires enhanced oil recovery (EOR) techniques. The development method and expected recovery gain depend on economic, environmental, and technical considerations. Therefore, operating companies conduct comprehensive studies prior to field development to select the optimal production method. This study focused on developing the Umm Niqa oil field located in the north of the state of Kuwait. A reservoir screening study for the target formation evaluated the different EOR techniques followed by experimental core flooding, which was the main methodology utilized to achieve the study objectives. Water flooding tests and steam injection were conducted using real core plugs of unconsolidated sandstone. However, the repeatability of the tests using the field core plugs failed; therefore, the study was completed using synthetic core plugs. The operating company provided the heavy oil samples. Water flooding runs were conducted at 33, 66, and 93 °C, while the steam injection was performed at 160 and 315 °C. The oil recovery results revealed that steam injection at 315 °C was the most suitable and optimal method to achieve production targets.

395a Polyazeotropy for Binary Mixtures: Methanol and Diethylamine

Nooram Anjum, Shigeo Oba, Toshihiko Hiaki, Chau-Chyun Chen

Azeotrope is a mixture of liquids which boil at a constant temperature at a given pressure. Although such mixtures are difficult to separate, they have proven useful to overcome various challenges such as minimization and recovery of waste of chemical processes, thus helping build a greener world. Therefore, it renders development of a thermodynamic model to correlate vapor liquid equilibrium conditions for mixtures exhibiting such behavior[1]. In general, one stable azeotrope formation occurs for a mixture for a given VLE conditions. The property of a mixture that produces more than one stable azeotrope at a given temperature or pressure is known as polyazeotropy. Representing the VLE data for a mixture exhibiting polyazeotropy has been of interest for many years.

The present study focuses on the binary mixture consisting of Methanol and Diethylamine. This mixture involves self-association between the components as well as cross association which is why mixed deviations from Raoult's law are observed. This behavior is not observed at low pressure for a constant temperature, but as the pressure increases double azeotropy can be detected. This study includes the analysis of association extents (bounded or non-bounded sites) to explain how the double azeotropes are related to the associations. Correlating the VLE data using UNIQUAC/UNIFAC model is insufficient as it only works for low and moderate pressures[2]. Since the system exhibits cross and self-association, Non-random two-liquid (NRTL) model is also not well suited as it is does not account for specific chemical associations[3]. Therefore, in order to represent the given system, Non-random two-liquid activity coefficient model with association theory[4] is used. This model combines Wertheim's perturbation theory for associations between the constituents with the classical NRTL model accounting for the non-idealities in the liquid phase. It regresses binary interaction parameters and unary molecule specific parameters to provide a satisfactory correlation for the experimental VLE data obtained from literature and successfully correlate the VLE data for the given binary mixture for a wide range of pressures.

References

1. Wisniak, Jaime, Hugo Segura, and Ricardo Reich. 'Polyazeotropy in binary systems.' Industrial andamp; engineering chemistry research 35, no. 10 (1996): 3742-3758.

2. Aucejo, Antonio, Sonia Loras, Rosa Muñoz, Jaime Wisniak, and Hugo Segura. 'Phase equilibria and multiple azeotropy in the associating system methanol+ diethylamine.' Journal of Chemical andamp; Engineering Data 42, no. 6 (1997): 1201-1207.

3. Bala, Aseel M., and Carl T. Lira. 'Relation of Wertheim association constants to concentration-based equilibrium constants for mixtures with chain-forming components.' Fluid Phase Equilibria 430 (2016): 47-56.

4. Hao, Yifan, and Chau-Chyun Chen. 'Nonrandom two-liquid activity coefficient model with association theory.' AIChE Journal 67, no. 1 (2021): e17061.

395b The Investigation of the Effects of Asphaltene Molecular Properties on Asphaltene-Solids Heterogeneous Interaction

Weiyi Kong, Rizwanur Rahman, Michael P. Hoepfner

Asphaltenes are a class of large aromatic organic compounds in crude oil that are soluble in aromatic hydrocarbons like toluene but insoluble in n-alkanes like n-heptane. Asphaltenes are a high polydispersity system diverse in an aliphatic chain(s), aromatic core size, and heteroatom(s). Some oils may contain varying concentrations of inorganic solids depending on the source locations. Small solids (< 1 micron) are difficult to remove during gravimetric and centrifugal separations are commonly performed to extract bitumen from oil sands. Inorganic solids presenting crude oil promote asphaltenes precipitation rate. Moreover, the insoluble asphaltenes assist in bridging inorganic solids into large flocs However, it is not well-understood which type of asphaltenes and why they interact with solids. The primary objective of this research is to correlate the asphaltene molecular structures to the interactions with solids during the asphaltene phase transition and investigate the mechanism of heterogeneous interaction. We hypothesize that the asphaltene-solid heterogeneous interaction is a multi-component phase transition process, and solids presenting might impact asphaltene thermodynamic properties. Asphaltenes will be fractionated based on whether they participated in heterogeneous interaction and then characterized their molecular properties, including double bond equivalent (DBE), heteroatom content, and molecular weight (carbon number). We plan to apply Fourier Transform Ion Cyclotron Resonance (FT-ICR) Mass Spectrometry (MS) with a 9.4 Tesla magnetic field to characterize asphaltenes molecular features. We will use principal component analysis (PCA) to correlate the molecular features promoting asphaltene-solid heterogeneous interaction. The cumulative knowledge compiled through the high-resolution mass spectrometry, scattering experiments, and principal component analysis should provide better insight into the driving force of asphaltene-solid heterogeneous interaction.

395c The Ideal Gas Contribution to Liquid Heat Capacity

Cassandra Guffey, Joseph Bloxham, Neil Giles, Thomas Knotts IV, Wade Vincent Wilding

Liquid heat capacity (LCP) can be thermodynamically described as the sum of the ideal gas heat capacity (ICP) and residual heat capacity. To examine the contribution of ICP to LCP, the DIPPR database was searched for all experimental liquid heat capacity data points. For each of these points, the corresponding ideal gas heat capacity was calculated using the accepted ICP correlation at the experimental temperature. The resulting test set for this work contained approximately 1100 unique compounds. Chemical family studies were performed to show the relationship between ICP and LCP. The results of these studies will be a useful tool as quality checks in the DIPPR compound evaluation process. The ICP-LCP relationship may also serve as a basis for the creation of future LCP prediction methods for a wide range of chemical families.

395d Modeling of Nitrogen + Fuel Phase Behavior Using the Helmholtz-Energy-Explicit Equation of State

Aaron Rowane, Ian H. Bell

The commercial jet fuel market is expected to grow from 106 billion gallons to 230 billion gallons by 2050 globally. Unlike for light-duty vehicles, electrification of the aviation industry is not expected to happen in the near future. Thus, sustainable aviation fuel (SAF) remains a promising, renewable, alternative to petroleum derived fuel. SAF can be produced biowaste and is made up of chemical species mirroring those found in petroleum derived fuels including *n*-paraffins, *iso*-paraffins, and naphthenes (ring/cyclic compounds). The extreme conditions the fuel is exposed to throughout the injection and combustion process are pressures up to 100 MPa and temperatures exceeding 700 K. High fidelity equations of state (EoS) that apply to SAF to enhance the design of jet engines using computational fluid dynamics are needed. Currently, REFPROP (version 10.0) includes multiparameter pure fluid EoS for several compounds included in SAF. However, the appropriate algorithms and mixture models needed to study mixtures of fuel-nitrogen (as a surrogate for air) are not currently available in REFPROP. In this study we employ the Templated EQuation of state Package (teqp) to model fuel-nitrogen mixtures using the Helmholtz-energy explicit EoS. The performance of the Helmoholtz-energy-explicit EoS is compared to existing models previously used to model fuel-nitrogen mixtures such as the PC-SAFT EoS. The overarching goal of this project is to assess the performance of the Helmholtz-energy-explicit model for its eventual incorporation into future versions of REFPROP.

395e Improving the Capability of Process Simulators to Represent Solid-Fluid Equilibria Applied to Natural Gas Liquefaction: The Methane + Neopentane System at Low Temperature

David Bluck, Seiya Hirohama, Nevin Gerek Ince, Marco Campestrini, Paolo Stringari, Freddy Garcia, Jean-Jacques Bartuel, Gang Xu

Natural gas is mainly methane, but it contains a number of other components, such as linear, branched, and cyclic hydrocarbons, aromatics, sulphur compounds, carbon dioxide, water, nitrogen, etc. Natural gas can be transported in pipelines or in its liquefied form, LNG, depending on the distance between the production site and the place of its utilization, the existence of geographical barriers, geopolitical considerations, or the need of a more flexible market. In order to be liquefied, natural gas must be cooled to about 110 K. This temperature is achieved after the natural gas is cooled in the Main Cryogenic Heat Exchanger (MCHE) at high pressure, usually higher than 50 bar, and then expanded down to the atmospheric pressure. The triple point temperatures of most of the natural gas components are higher than the temperatures that natural gas achieves in the MCHE. Consequently, if the mole fraction of these components in natural gas is higher than their solubility limits at the temperature and pressure in the MCHE, they can form solid phases in the heat exchanger. The solid formation increases the pressure drop in the MCHE and reduces the heat transfer efficiency. This means more frequent maintenance operations are required, that can have a very important negative impact on the profitability of the liquefaction plant. In order to avoid solid formation in the MCHE, an accurate knowledge of the solid-liquid, solid-vapor, and solid-liquid-vapor equilibria is mandatory to design purification units allowing reducing the content of the potential solid formers below their solubility limits before entering the MCHE. For such a purpose, the capability of the process simulators of representing accurately the phase diagram of natural gas is very important to avoid designing oversized purification units or to limit the energy and material consumption in these units.

This work is the result of a collaboration among a natural gas producer, an EPC (Engineering, Procurement, and Construction) company, a simulation software company and a research laboratory with the aim of improving the knowledge of the solubility limits of some key impurities in LNG. In this perspective, neopentane has been selected as a potential solid former because of its high triple point temperature, 256.6 K. In addition, the only two articles in the literature reporting solubility data of neopentane in methane (the main natural gas component) are not in agreement each other. A wide experimental investigation of the liquid-vapor, solid-vapor, solid-liquid, and solid-liquid-vapor equilibria for the methane + neopentane system at low temperature has been carried out and presented in Campestrini et al. 2022. This work shows how the data produced by the research laboratory are made available to industries thanks to the implementation of dedicated models and algorithms in a process simulator. In particular, an accurate model, based on the Soave-Redlich-Kwong (SRK) cubic Equation of State (EoS), has been developed for the phase equilibria of the methane + neopentane system including solid phases. In addition, a robust multi-phase flash algorithm has been developed for the solid-liquid(-liquid)-vapor equilibrium calculation. Rigorous thermodynamic phase stability analysis is used to determine the correct number of phases that will coexist at any given temperature and pressure. The new model and the phase equilibrium algorithm are integrated in the process simulator to improve the accuracy in the representation of the phase boundaries in natural gas and to allow an optimized design of the liquefaction units.

395f Methane Mass Transfer in Liquid Hydrocarbon Mixtures – Measurement and Modeling in Synthetic Porous Media

Sheng Hu, Theodore Tsotsis, Kristian Jessen

Methane (CH_4) dissolution and mass transport in liquid hydrocarbon mixtures is of key interest in enhanced recovery from tight shales. In this research, we study experimentally CH_4 mass transport through synthetic model porous media saturated with liquid hydrocarbons, with the objective being to quantitatively investigate the effect of capillarity and pore structural characteristics on methane transport through such media. The overarching goal of our research is to gain a better fundamental understanding of the mechanism of methane dissolution and transport under nanoconfinement.

In our studies, we employ a high pressure and temperature Wicke-Kallenbach diffusion cell. Vycor 7930 and AGC-40 (*Advanced Glass andamp; Ceramics, LLC*) porous glass with average pore size of 7 and 4 nm, respectively, are used as the model synthetic porous media. By monitoring the pressure drop in the bulk gas phase, we can measure the rate of dissolution and diffusion of CH_4 through the model porous silicas.

The phase behavior and transport properties of methane/hydrocarbon systems in the bulk have been well studied by our Group and others. However, the nanoconfinement of these fluid mixtures in tight rocks causes their physical behavior to deviate from the corresponding bulk phase behavior. In our study, we have developed a model that considers the impact of pressure difference across the vapor-liquid interface and of liquid swelling due to the methane dissolution in the pore space. The results of our study to date show that this model can predict the mass transfer behavior of methane/n-alkane systems in nano-confinement from the bulk-phase information.

395g Exploring Liquid Crystalline Phase Behavior of Pyrene Pitch Oligomers Isolated Via Supercritical Extraction

Graham W. Tindall, Jefferson T. Giles II, Adam S. Gross, Yunlong Zhang, Amy C. Clingenpeel, Kazem V. Edmond, Manesh Gopinadhan, Stuart E. Smith, David Perkins, Jonathan D. Saathoff, Mark C. Thies

Discovered in the mid-1960s, liquid crystalline (also known as mesogenic/mesophase) pitches became the foundation for a new class of highly graphitic, carbon-based materials. Due to their high carbon content and molecular order in the liquid phase, mesogenic pitches can be melt-processed into carbon materials that possess a valuable combination of mechanical strength, modulus, thermal conductivity, and temperature resilience. However, not all mesophase pitch (MP) precursors are made equal and they must be optimized based on processing capabilities. For instance, higher amounts of mesophase may yield a better final product, but also make the melt more prone to fracturing while being drawn. Furthermore, the softening point of the MP must be low enough so that it will flow before thermal degradation becomes significant, but if the softening point is too low, the MP can undesirably liquefy in downstream processing, wiping out any macroscale shaping. Thus, a thorough understanding of the pitch precursor and its properties/phase-behavior is necessary for successful processing.

Although an improved fundamental understanding of petroleum-based pitches is of great interest, investigations are severely hindered by the innumerable array of individual compounds that comprise these pitches. As a model system, pyrene pitch is ideal for two reasons: 1) the relatively low extent of fragmentation means that pyrene pitches are chiefly composed of only a handful of oligomers, and 2) pure pyrene trimer (MW = 598 Da) forms 100% mesophase with a melting point below 300°C.

Supercritical extraction has proven to be a valuable tool for investigating the properties and phase behavior of pyrene pitches, as it is the only known tool capable of isolating individual pitch oligomers on at least a gram-scale. With these oligomers isolated, they can then be recombined into mixtures of well-defined composition, and the mesophase/isotropic phase behavior of these mixtures characterized. This allows for a systematic exploration of phase behavior, where each "designed mixture" is characterized by the mesophase content (as a function of temperature), softening point, and char yield. Because smaller oligomers depress and larger oligomers elevate the softening point of a mixture, understanding their impact is critical for successfully optimizing the properties and mesogenicity of a pyrene pitch precursor.

599a Solutions of Nonionic Surfactants in CO_2 to Improve Unconventional CO_2-EOR Via Wettability Alteration

Lauren Burrows, Foad Haeri, Deepak Tapriyal, Sean Sanguinito, Parth Shah, Peter Lemaire III, Robert M. Enick, Dustin Crandall, Angela Goodman

CO_2 huff n' puff is a promising method for enhanced oil recovery (EOR) in unconventional shale formations. In this work, we postulate that CO_2-EOR may be improved by the dissolution of surfactants into CO_2 if both oil-rich and CO_2-rich phases are present. A complete pressure-composition phase behavior diagram of CO_2-dead Eagle Ford crude mixtures indicates that at 80 °C and pressures as high as 62 MPa, two such phases co-exist for CO_2-rich mixtures. Therefore the dissolution of a surfactant in CO_2 has the potential to improve oil recovery via wettability alteration from oil-wet to CO_2-wet and/or CO_2-oil interfacial tension (IFT) reduction. Three nonionic surfactants (linear dodecyl ethoxylate Indorama SURFONIC® L12-6, branched nonyl ethoxylate Indorama SURFONIC®TDA-9, and branched nonylphenol ethoxylate Indorama SURFONIC® N-100) were evaluated for CO_2 solubility, shale wettability alteration, ability to increase oil recovery during laboratory huff n' puff experiments, and effect on CO_2-oil interfacial tension (IFT). Each surfactant exhibited CO_2-solubility at temperatures up to 100 °C of roughly 1 wt% at pressures commensurate with CO_2-EOR. The presence of surfactant in CO_2 did not have a significant effect on CO_2-oil IFT. High-pressure, high-temperature contact angle experiments demonstrated that the (CO_2+surfactant) solutions can shift the wettability of oil-wet shale toward CO_2-wet and water-wet. CO_2 and (CO_2+surfactant) huff n' puff experiments for oil-saturated Eagle Ford, Mancos, and Bakken shale cores immersed in CO_2 showed that CO_2 solutions containing the branched nonyl ethoxylated alcohol gave highest cumulative oil recoveries of 72% and 75% (at 0.01 wt% and 0.1 wt% of TDA-9, respectively) compared to 71% recovery for the core immersed in pure CO_2. Analysis of oil produced during confined-core huff n' puff experiments indicated that heavier oil components were produced when the surfactant was added to CO_2. These results suggest that CO_2-soluble surfactants may increase oil recovery by wettability alteration rather than IFT reduction.

599e Production Performance of US Oil Shale Plays Related to Their Depth Profile and Initial Gas Oil Ratio (GOR)

Rasoul Sorkhabi, Palash Panja

Compared to the conventional oil plays, the shale revolution is relatively a recent development and our understanding of production sweetspots related to petroleum system characterization is a new frontier for research. The objective of this study is to assess the production performances of eight major shale plays in the US basins related to payzone depth and initial gas oil ratio (GOR). We collected well information and production data for 35,909 horizontal wells with at least 24-month production history for the following plays: Eagle Ford (10,782 wells), Austin Chalk (1014), Permian Wolfcamp (3,543), Permian Bone Spring (1726), Woodford (712), Barnet (733), Denver Niobrara (4056), and Bakken (13,343). For consistency in comparison, we calculated normalized production (STB per foot per month) and plotted them against payzone depths. We also investigated the relationship between the initial GOR and the normalized production. These parameters were selected to understand the production performance of the plays in relation to geologic and reservoir conditions. Our data analytics offers significant insights into the performance behavior of the shale plays. Although each shale play has its own characteristics, some general trends are observed. Overall, production per depth follows a bell-shaped profile that can be divided into three zones. The Upper tail at depths of 400-6000 ft. are too shallow for oil generation, and we interpret oil accumulations at these shallow payzones as intra-formation migrated oil from deeper levels. The bulges of the bell curve, which contains the majority of oil wells, are located at depths of 6000-12000 ft; they exhibit the best performance with maximum production per unit of STB/ft//month. These wells and payzone are situated in the present-day oil generation window (60-120 °C). The lower tail of the bell curve at depth of >12,000 feet zone enters the gas window and is thus less productive of oil. The maximum production is partly influenced by fluid properties like initial GOR. Our analysis for all eight plays finds an inverse relationship between normalized oil production and increasing values of initial GORs. A frontier was drawn to obtain the maximum limit of production for a given initial GOR. The novel aspect of our study is reverse engineering of production data to petroleum systems and geological characterization of the shale plays as self-sourced reservoirs. We note that (1) there is intraformational migration of oil to shallower levels; (2) the best producing payzones are situated within the present-day oil windows: and (3) payzones with initially lower GORs have better production performance through the lifespan of each well.

599f A Study of Capillary Condensation Phenomena in Mesoporous Materials

Jiyue Wu, Sheng Hu, Kristian Jessen, Theodore Tsotsis

The phase behavior of a fluid can be significantly affected by its confinement in nanopores. Within mesoporous materials, the strong fluid-solid interaction in the pores can induce a vapor to liquid phase change that occurs below the saturation pressure of the bulk fluid, a phenomenon which is known as capillary condensation. Identifying the conditions for which capillary condensation takes place in the subsurface under reservoir conditions plays an essential role in the design of unconventional oil and natural gas (UOG) resource systems. The focus of this study is the study of the phase behavior of low molecular weight (MW) alkanes confined in nanopores under reservoir relevant conditions and examining whether or not capillary condensation takes place.

For the study, we employ model mesoporous silicas, both amorphous and crystalline. We have selected for our study ethane as a model low MW hydrocarbon component of shale gas. We utilize a high-pressure thermogravimetric microbalance to study the adsorption and potential capillary condensation of ethane in the silica nanopores, as well as a volumetric system under cryogenic conditions. The experiments are complemented with models of the phase behavior, including the classical Kelvin Equation (KE), and a reformulated KE model based on the Peng-Robinson Equation of State (EOS). They are complemented by Grand Canonical Monte Carlo (GCMC) simulations to help reveal the possible mechanisms behind the fluid-solid interaction within the pores of these materials.

396a Towards Rational Design of Metal-Embedded Zeolite Catalyst for Improved Catalytic Performance in Microwave-Assisted Methane Dehydro-Aromatization

Sanjana Karpe, Xinwei Bai, Brandon Robinson, Jianli Hu, Goetz Veser

Due to the limitations in the transportation of abundantly available natural gas to point-of-use locations, large amount of natural gas is flared causing significant environmental pollution. This motivates interest to convert natural gas to easily transportable value-added liquids. One such conversion route is the direct conversion of methane to benzene via methane dehydro-aromatization (MDA). Typically, metal/zeolite catalysts are used for this reaction, where methane is activated on a metal site, oligomerized, and further aromatization takes place at the Bronstedt acid sites (BAS) of the zeolite where the zeolite provides shape selectivity towards the desired product benzene. However, coking and lengthy catalyst induction times are major challenges for this reaction.

Microwave-assisted heterogeneous catalysis may offer a way to mitigate these problems by selectively heating the solid catalyst, thus reducing unwanted side-reactions in the—comparatively cold—gas phase, reducing coking, and accelerating catalyst induction times. We had previously identified Fe-ZSM-5 as a promising catalyst for microwave-assisted MDA. Fe-ZSM-5 showed strongly increased methane conversion and shortened induction time under microwave catalysis conditions. However, increased conversion was obtained at the cost of lower selectivity to the desired product benzene. The results could be explained by the formation of strong metal hot spots which lead to deep dehydrogenation of methane to coke, while the relative weak microwave susceptibility – and hence low temperature - of the zeolite would result in low activity of the BAS on the zeolite and hence in poor aromatics selectivity.

Based on these results, in the present work we are exploring the design of SiC@Fe-ZSM-5 core@shell catalysts to reduce these intra-catalyst temperature gradients. Goal of the catalyst design is to combine the activity of the metal-exchanged zeolite with the well-known high microwave susceptibility of SiC while keeping the two components in intimate contact and hence maximizing heat transfer during microwave irradiation. By directly growing the zeolite phase onto a SiC, which is much more strongly heated by microwave irradiation, we aim to counter the low microwave absorption of the zeolite phase and hence reduce the intensity of the metal hot spot.

SiC@Fe-ZSM-5 core@shell catalysts with well-defined zeolite shells were successfully synthesized and characterized, and then evaluated in MDA at conventional thermos-catalytic and microwave-assisted conditions. Catalytic performance was evaluated in comparison between different catalyst configurations, including pure Fe-ZSM-5 and a corresponding physical mixture of SiC and Fe-ZSM-5 as a reference. We find indeed that the core@shell catalyst has improved catalytic performance in comparison to a pure Fe-ZSM-5 catalyst and the physical mixture, in agreement with the hypothesis that this configuration would reduce the intensity of the metal hot spot. Differences in reactive performance of the different catalyst configurations (conversion, selectivity, and induction time) can be explained consistently based on expected temperatures of the active sites (metal vs BAS in zeolite) and the intensity of metal hot spot attained in these catalysts. Ongoing work is extending these studies onto further tuning of the intensity of metal hot spots by tuning the core@shell structure, i.e. changing the relative size of SiC core vs Fe-ZSM-5 shell and thus changing the intensity/effectiveness of the indirect heating via the SiC core.

Overall, our work proposes an effective way to control internal catalyst temperature distribution under microwave-assisted catalytic conditions via rational catalyst design. We expect that such an intentional design of catalysts for use in microwave-assisted heterogeneous catalysis will enable taking full advantage of this promising reactor concept.

396b Deactivation Study of Mo-HZSM-5 Catalyst for Microwave-Assisted Methane Aromatization: Effect of Reaction Parameters

Ashraf Abedin, Pranjali Muley, Victor Abdelsayed, Xinwei Bai, Hari Paudel, Daniel Haynes

Associated gas is a by-product of oil production and is most often flared due to transport limitations in the existing pipeline infrastructure[1]. This waste of an important domestic resource can be minimized by using the natural gas as a source for the synthesis of valuable, energy dense products such as benzene. For modular conversion of natural gas at the well-head, the use of modular microwave reactors can include rapid heating for on-demand chemical production, electrification for use of renewables and high heat fluxes in small volumes for process intensification benefits in compact reactors.

Methane is known as the primary constituent of natural gas. Direct catalytic conversion of methane to benzene using methane aromatization (MDHA) can convert methane directly to benzene and hydrogen[2] (eqn. 1).

$6CH_4 = 4C_6H_6 + 9H_2$; (ΔG_{298K} = +433 kJ/mol, ΔH_{298K} = +530 kJ/mol) (1)

Very high temperatures (700°C-1000°C) are needed to drive the equilibrium towards the formation of benzene and hydrogen. However, challenges of rapid catalyst coking along with thermodynamic limitations represent a major challenge towards a process with high conversions and product yields[3]. Mo/HZSM-5 catalyst has been demonstrated as the most active/selective catalyst for MDHA due the proximity of molybdenum to internal Bronsted acid sites for C-C bond formation and its shape selective pore structure to enable benzene formation. Microwave assisted activation allows localized heating on the active sites while reducing the bulk reaction temperature, which can alter side reactions that produce carbon deposition on this catalyst[4]. In this work, the correlations between catalyst activity and reaction parameters under microwave are investigated to determine the activity-structure relationship and the role of carbon under microwave heating.

Mo/HZSM-5 catalyst with 4% Mo loadings was used in this study. For even distribution of heating within the catalyst bed, a silicon carbide (SiC) powder was physically mixed with the catalyst to improve bed heating and temperature uniformity. The catalyst was thermally activated under CO, followed by MDHA reaction while varying temperature, WHSV and methane concentration via microwave-assisted heating. Different characterization tools were used to study its structure, carbon formation and activity such as BET, TPO, XANES, TGA and XRD techniques.

For all runs under microwave, the catalyst was highly selective to benzene and ethylene as primary reaction products, with little selectivity to toluene or xylene. When Mo-ZSM5 was thermally reacted at 700^0C, a lower yield of benzene was observed in comparison to microwave reaction at 500^0C. L_{III} edge XANES showed higher Mo oxide reduction under microwave compared to thermal heating, suggesting faster activation of Mo under microwave. This maybe due to the selective heating of Mo active sites compared to ZSM-5 sites, thus producing higher benzene yields at lower bulk temperature.

Carbon formed during microwave runs between 400-700^0C and analyzed by thermogravimetric analysis (TGA) showed the highest accumulation of carbon at 700^0C (figure 1). BET results indicate a reduction in pore volume, which suggest the carbon is depositing in the mouth of the pores. At 700^0C in microwave, this results in maximum pore blockage, which restricts the reaction intermediates to access ZSM-5 pores and form aromatics. For temperatures in the range of 500-600^0C, the maximum production of benzene can be correlated to polymeric carbon formation observed in TPO. Carbon formation increased with decreasing space velocity(figure 1), as higher gas-solid contact period caused additional carbon deposition in the catalyst. With decreasing space velocity, a transition of carbon formation was observed from polymeric to graphitic region. At 500^0C, which is the optimum microwave temperature for higher benzene yield, an increasing amount of carbon was formed for all runs when methane concentration was increased from 10% to 70%(figure 1). Carbon deposition increased and BET surface area decreased when methane concentration increased, indicating higher catalytic activity. This decrease in surface area could be related to higher local temperatures as carbon is an excellent microwave absorber. Deposition of carbon increased with temperature, gas-solid contact period and methane concentration thus enhanced microwave heating and resulted in higher catalytic activity. Compared to microwave reaction at 500^0C, thermal reaction at 700^0C resulted in lower carbon deposition, which led to deactivation. This indicates the role of the deposited carbon as a possible active site for promoting microwave assisted MDHA. However, a significant temperature gradient exists within the catalyst bed, with core of the catalyst bed reaching a much higher temperature than the surface in microwave, which suggests that in microwave, MDHA can occur at much higher temperature compared to thermal reactions. The study indicates that under traditional heating carbon formation deactivates the active sites, however, under microwaves, the presence of localized carbon plays an important role in coupling microwave energy and leading to localized heating that improves selectivity to benzene.

Keywords: microwave dehydroaromatization, catalyst characterization, microwave deactivation.

References

1. Mac Kinnon, M.A., J. Brouwer, and S. Samuelsen, *The role of natural gas and its infrastructure in mitigating greenhouse gas emissions, improving regional air quality, and renewable resource integration.* Progress in Energy and Combustion Science, 2018. **64**: p. 62-92.
2. Abedin, M.A., et al., *Sulfated hafnia as a support for Mo oxide: A novel catalyst for methane dehydroaromatization.* Catalysis Today, 2020. **343**: p. 8-17.
3. Abedin, M.A., et al., *Methane dehydroaromatization using Mo supported on sulfated zirconia catalyst: Effect of promoters.* Catalysis Today, 2021. **365**: p. 71-79.
4. Abdelsayed, V., D. Shekhawat, and R.S. Tempke, *Zeolites interactions with microwaves during methane non-oxidative coupling.* Catalysis Today, 2021. **365**: p. 88-102.

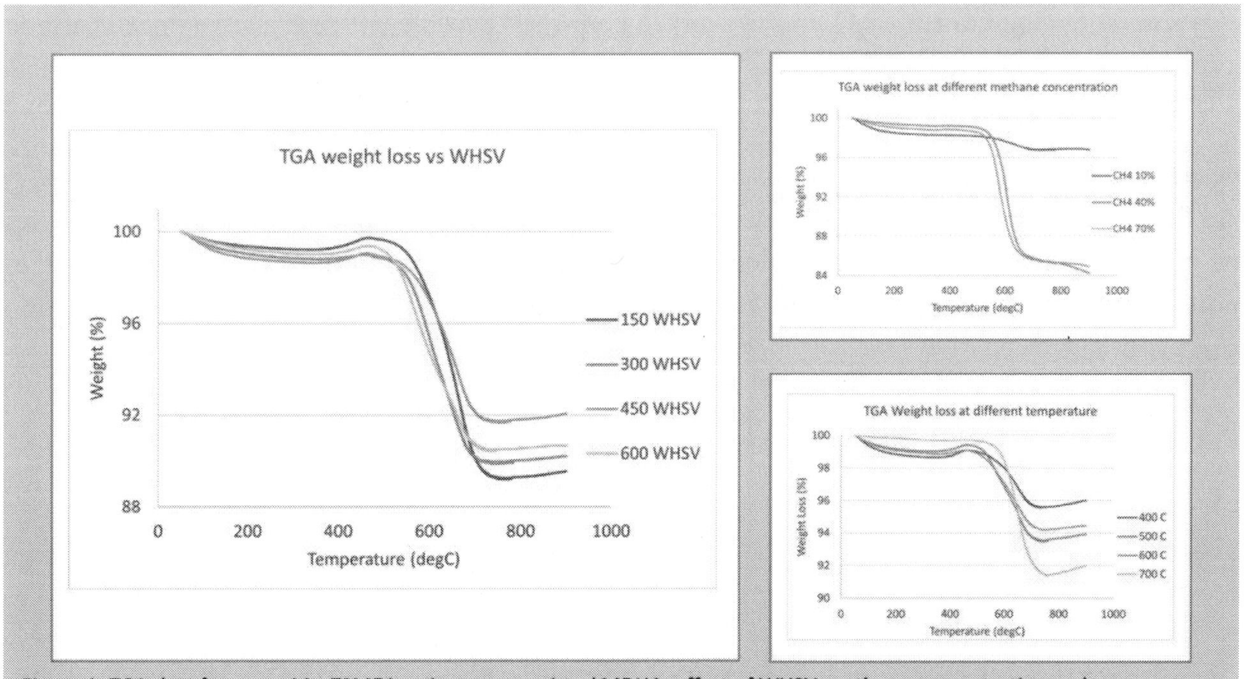

Figure 1: TGA data for spent Mo-ZSM5 in microwave assisted MDHA- effect of WHSV, methane concentration and temperature

396c Ethane Dehydroaromatization with CO_2 Co-Feed over Microwave Synthesized Core-Shell Transition-Metal-Carbide@ZSM-5 Material

Ashley Caiola, Brandon Robinson, Sean Brown, Jianli Hu

The continuous demand for cleaner fuel alternatives drives the world's search for substitute fuel sources. Natural gas has been a promising fuel source and currently makes up a large portion of the North American energy production. The conversion of natural gas to value-added chemicals, typically involves the indirect production of synthesis gas (syngas) as an intermediate before the production of the desired product. Steam reforming and partial oxidation are two of the most common indirect methods of natural gas conversion. Both involve the intermediate synthesis of syngas, which is both energy and capital-intensive. Alternatively, the direct conversion of light alkanes, such as methane and ethane, into aromatics eliminates the expensive intermediate production of syngas and reduces greenhouse gas emissions. Although the conversion of natural gas by the direct method shows promise as a cost-efficient approach, the commercialization of this process faces some key technical challenges. Rapid catalyst deactivation, low selectivity to desired products, and long regeneration times are the current challenges preventing the commercialization of dehydroaromatization (DHA). The accumulation of carbon on the catalyst surface, with time on stream (TOS), can block the active sites resulting in a decrease in the selectivity and deactivation of the catalyst over time. Regeneration of the catalyst is typically carried out in oxidative atmospheres, where an exothermic reaction burns the coke off the catalyst thus producing CO_2. The addition of oxygen species, such as O_2, CO, and/or CO_2 to the inlet gas as a co-feed or pulse would help to avoid this rapid deactivation of the catalyst.

The use of CO_2 as a soft oxidant for ODHE and ODHA comes with several advantages which consist of reduced coke formation, prevention of over oxidation of metal catalyst, and through the enhanced equilibrium conversion obtained due to the removal of hydrogen via the water gas shift mechanism. Also, the use of CO_2 as an oxidant can have tremendous impacts on CO_2 emissions by consuming CO_2 and producing CO and H_2O. In this work, a core-shell particle with a transition metal carbide core and a zeolitic shell, H-ZSM-5, was synthesized using a microwave synthesis unit for the tandem ethane aromatization and dehydrogenation with CO_2 as the oxidant. The use of ethane allows for both the simultaneous reactions of ODHE and DHA at temperatures of 600°C. The DHA catalyst support, ZSM-5, was synthesized via microwave irradiation and metal promoters used consist of Mo, Fe, Ga, and Pt. The transition metal carbide catalyst core, TiC or Mo_2C, promotes the formation of ethylene, the increased ethylene concentration along with the production of CO from CO_2 can allow for the more efficient production of aromatic. Using CO_2 as a soft oxidant could also allow for extended reaction times.

396d Enhanced Dry Reforming of Methane Using Pseudo Catalytic Metal Oxide and Nanoparticles

Pinak Mohapatra, Qichang Meng, Anuj Joshi, Sonu Kumar, Patricia Loughney, Ashin Sunny, Lang Qin, Zhuo Cheng, Vicky Doan-Nguyen, Liang-Shih Fan

As the world is rapidly developing alongside an increasing population, energy demands have risen dramatically. Historically, technological advances have improved human well-being, but many have also unintentionally contributed to environmental degradation and damage to society because of harmful greenhouse gas generation such as CO_2. A significant contribution to this has been the use of fossil fuels over the past two centuries. To ensure that engineering makes net-positive contributions to sustainable and economic process development, especially for reforming fossil fuels, we need approaches to develop technologies that reduce the chance of such unintended harm. Due to fracking or hydraulic fracturing, there has been a steep rise in natural gas production. Large amounts of natural gas are flared because of storage limitations, which leads to energy wastage and engenders a considerable amount of CO_2. Materials that can convert the flared natural gas containing primarily CH_4 into synthesis gas ($CO+H_2$) by employing CO_2 as the reforming gas have lately received pressing attention. Through this process, the two greenhouse gases CH_4 and CO_2 are utilized, and this reaction is classified as Dry Reforming of Methane (DRM). However, DRM at high temperature where the reactants exhibit maximum conversions faces two major hurdles concerning its catalyst: 1) Carbon deposition on the active metal sites and 2) Sintering of the active metal, which reduces the activity. Several researchers have proposed ways to mitigate these issues, including alternative synthesis techniques, mesoporous supports to prevent agglomeration of the active metal, injection of steam to reduce the amount of carbon, etc. Though, these strategies increase the life of the catalyst but haven't solved the issue. Consequently, the catalysts fail before 200 hours of operation. Based on our thermodynamic calculations, efforts to curb carbon deposition under typical high-temperature conditions is thermodynamically unattainable.

To address the issue, a higher proportion of CO_2 with respect to methane is investigated. At this ratio, carbon deposition is thermodynamically zero at pressures up to 30 atm, which means the process can be operated even at higher pressures, thus eliminating the enormous compression and decompression costs. A pseudo-catalytic metal-oxide (PMO) prepared using industrial solid state synthesis method is developed to process the higher CO_2 to CH_4 ratio. The pseudo-catalytic nature of the metal oxide is imparted by the PMO's tendency to participate in the reaction, forming stable reaction intermediates. The PMO material employs a unique syngas generation mechanism at high temperatures to efficiently convert CH_4 and CO_2 to syngas. Based on the demand, this high-quality syngas can be further converted into liquid fuels such as C3+ olefins and aromatics. Continuous testing at 1000C at CO_2/CH_4 ratio of 1.5 exhibited no carbon deposition for more than 850 hours. Further, the process showed 80% CO_2 conversion and near 100% CH_4 conversion during the testing. The syngas yield per mole of CH_4 is 3.7 versus the thermodynamic limit at 3.8. Besides, the quality of syngas obtained is $H_2:CO = 0.8$. Steam can be added to the reactants to alter the syngas quality further.

Furthermore, smaller size nanoparticles were also studied to investigate the effect of particle size on the reaction performance. The material is based on $FeNi_3$ nanoparticles supported by SBA-15 (FeNi3@SBA-15). The pore size and pore connectivity of SBA-15 have been tuned to achieve the optimal molecular diffusivity and reactivity. This tuning was achieved by controlling intermicellular interactions, through the choice of surfactant and the synthesis temperature. The FeNi3@SBA-15 samples were synthesized using wet impregnation method and their physicochemical characteristics were analyzed with Brunauer-Emmet-Teller (BET), transmission electron microscopy (TEM), X-ray diffraction (XRD), and small-angel X-ray scattering (SAXS) analyses. Their catalytic activity was studied by temperature-programmed reaction. The FeNi3@SBA-15 with a pore radius of 18.7 Å and pore volume of 0.461 cc/g exhibited high reaction kinetics and over 99% methane conversion at 900 °C. The potentiality of these nanoparticles can be investigated further for long term and high temperature testing similar to industrial grade PMO. Enhanced DRM can be an effective route to utilize the two greenhouse gases, converting them into value-added products.

396e Low Temperature "Super-Equilibrium" Reforming of Methane through Chemical Looping

Luke Neal, Chongyan Ruan, Fanxing Li

The complexity and cost of reforming reactors is a major barrier to distributed, natural gas-to-liquids technology. Such schemes rely upon partial oxidation/autothermal reforming (POx), where a pure oxygen feed is reacted with methane to form a ~2:1 H_2:CO syngas. One of the more obvious limitations of the process is the cost and safety concerns of producing and using purified oxygen. We have previously shown that chemical looping reforming (CLR), in which the oxygen for the methane POx reaction is provided from a mixed-metal oxide, can greatly alleviate these issues. However, both traditional POx and CLR have a more subtle limitation for distributed scales: at lower temperatures (<750 °C), the equilibrium reaction favors the coexistence of significant amounts of CO_2, H_2O and CH_4. At 650 °C, for example, yield of syngas is limited to 57%. While increasing the oxygen stoichiometry will increase methane conversion, syngas yield is reduced, and CO_2 and H_2O yields increase. The inability to operate below 750 °C greatly reduced the types of materials reactors and valves can be made from, leading to high costs.

In this work, we present a "super-equilibrium" mode for CLR operation, which allows for avoidance of the thermodynamic limitations on reforming. In this configuration, two types of chemical looping catalyst are employed, a POx catalyst and a Splitting catalyst. A CLR POx catalyst allows for high conversion of methane at low temperatures (e.g. 650 °C), resulting in a mixture of syngas, CO_2, and H_2O. High methane conversion is achieved by using a stoichiometric excess of oxygen. To improve syngas yield, the CO_2 and H_2O are split to CO and H_2 through the oxidation of a splitting catalyst. Integration of the splitting and POx catalysts allows for pseudo continuous operation with "super-equilibrium" syngas yields. Here we describe the scheme in detail, including thermodynamic analysis of the overall reaction and ideal mixed oxide systems. Experimental work confirming "super-equilibrium" yields is also shown.

396f Microwave Catalytic Conversion of Acetylene for Co-Production of Hydrogen and Carbon Nanotubes

Sonit Balyan, Changle Jiang, Ashley Caiola, Jianli Hu

Single step natural gas conversion to value added product can drastically reduces the carbon footprint of chemical processing industries. Microwave plasma reactors (MWP) is an emerging process intensification for chemical industries to directly convert methane (CH_4), the major component of natural gas and offers benefits such as fast process dynamics, flexibility, high product yield and retards formation of unwanted by products during process intensification. MWP generated from CH_4 results to produce large amount of acetylene, which is mainly used in welding. To add value to acetylene from the downstream of methane plasma reactor, a novel approach towards formation of carbon nano tubes (CNT) and pure hydrogen (H_2) on bimetallic catalyst driven by microwave irradiation is studied. The preliminary data showed that the acetylene can be converted to carbon and hydrogen (Figure 1). The dominant gas product was hydrogen (~4 mol % concentration). Low concentrations of methane, ethane, and ethylene were observed in product stream. To further examine that the deposited carbon is in CNT form we have carried out acetylene pyrolysis at 400 °C over 10Ni-1Pd-Al_2O_3 (Figure 2) supported catalyst. From results it is verified that at 400 °C acetylene can be dehydrogenated with hydrogen as dominant gas with 3.2 mol% concentration in product stream at 150 minutes time on stream for 10Ni-1Pd-Al_2O_3 catalyst. We have further carried out characterization tools such as Raman spectroscopy, and TEM further access the property of carbon deposited over the catalyst during acetylene pyrolysis.

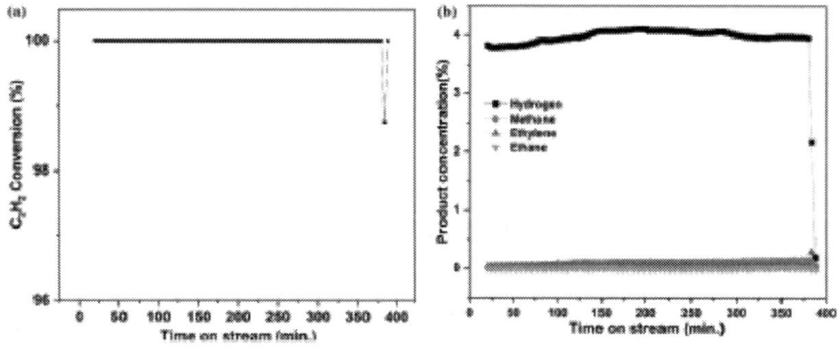

Figure 1. (a) Acetylene conversion, and (b) product concentration over 10Ni-1Pd-CNT at 500 °C

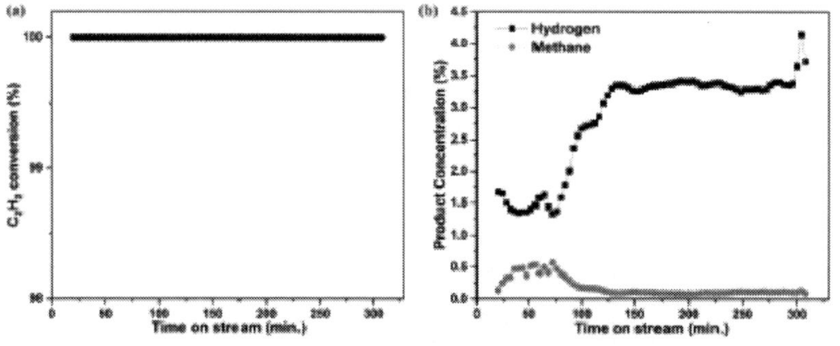

Figure 2. (a) Acetylene conversion, and (b) product concentration over 10Ni-1Pd-Al_2O_3 at 400 °C.

195a Building a Toolbox for Integrating Economic and Environmental Performance of Bio-Based Products

Samir Meramo, Peter Fantke, Sumesh Sukumara

Recent trends in biotechnology have shown accelerated growth of bio-based innovations, in the pharma, food and chemical industries. The resulting monetary influx has boosted initiatives towards developing new technologies to transform bio-based feedstocks into bio-based chemicals and derived materials. However, many biotechnological ventures remain resource-intensive (both in mass and energy), hampering the urgently needed progress of numerous biotechnological innovations to meet global sustainability targets and facilitate green transition toward sustainability. This is one of the main causes for what many initiatives are delayed reaching commercialization. Therefore, an opportunity of encouraging product development in the bio-based industry should consider analyzing the challenges from the inception, necessarily involving sustainability-driven designs at primary design phases to seek for potential improvement areas and optimize the technology earlier.

The innovations in biotechnology must ensure a transition toward sustainability by consistently advancing the economic and environmental performance of resulting bio-based products. Environmental sustainability assessment is a method implemented to estimate potential environmental impacts at different technology readiness levels. Among the impact assessment methods, Life Cycle Assessment (LCA) is a standardized tool to quantify environmental impacts over a product or technology life cycle. Two relevant applications of LCA are the analysis of the contributions of the life cycle stages to the overall performance, prioritizing improvements areas, and comparing different alternatives, to identify best-in-class solutions. Similarly, Techno-Economic Assessment (TEA) has been extensively applied in chemical process design for the technical and economic evaluation of processes considering financial aspects and cost parameters.

The application of TEA and LCA studies of bio-based products is not new. Still, there is a methodological limitation to effectively deal with the tradeoffs of the economic and environmental performance. Most of the studies have reported standalone TEA and LCA application, alluding to the need for a comprehensive framework that consistently integrates the economic and environmental performance. Such an instrument is necessary to guarantee a smoother transition of the biotechnology sectors towards overall sustainability. In addition, this methodology could guide the Randamp;D activities that are currently being explored by assessing the combined sustainability performance of bio-based initiatives. Previous studies on integration approaches reveal an absence of consistency regarding the defining criteria and methodological issues for integration. For example, LCA is a standardized methodology with four major phases: goal and scope, life cycle inventory, life cycle impact assessment, and interpretation. In contrast, TEA is a methodology that lacks a standard, so it could be performed differently by conserving the same principles (cost breakdown and financial, for example). Therefore, the integrated methodology should consider the described limitation to avoid lack of consistency. Alongside, to aid practitioners and researches in guiding the integrated use of LCA and TEA, a computer-aided tool is needed. Such a tool will also allow boosting sustainability-driven innovations in biotechnology.

A comprehensive toolbox that integrates LCA and TEA with application in the biotechnology sector will be presented in the present study. The toolbox analyzes bio-based products from a systems perspective, and life cycle thinking to avoid the burden-shifting and allows assessing the product from expanded boundaries to integrate its economic and environmental performance. Different computational tools are integrated to perform TEA and LCA, aligning systems boundaries, metrics, units, and uncertainty analysis. Firstly, mass and energy balances are generated in a process-modeling module through simulations. This allows access to inventories required to perform TEA and LCA. Then, specific TEA and LCA modules are executed to quantity and improve the economic and environmental sustainability performance. Finally, the results are combined in an integration module to illustrate the tradeoffs and provide an overall sustainability score.

A case study with application in biotechnology will be tested to show the functionalities of the toolbox. The case study will present a low maturity with a Technology Readiness Level between 1 and 2. A functional unit of 1 kg of bio-based product is set as a common function, to guarantee that inventory flows (and then the results) in LCA and TEA are taken under the same function.. The toolbox will illustrate the results in different modules, including LCA, TEA, and integrated dashboards. In the TEA dashboard, detailed economic results are illustrated, describing the cost breakdown and financial variables like internal rate of return, return on investment, and payback period. The LCA dashboard provides more information about the life cycle impacts and environmental hotspots considering midpoint and damage level results. Finally, the integration dashboard illustrates the combined results of the case study, integrating the TEA and LCA results, to arrive at a final score that allows benchmarking against existing or competing technologies the product system from a sustainability perspective.

195b Enhanced Sugar Yield from Cellulose By Ammonium Salt Based Pretreatment

Monidipa Bose, Eshita Saha, Shyamal Roy

The crystalline avicel cellulose was transformed into amorphous cellulose using aqueous ammonia and ammonium thiocyanate salt pretreatment. The pretreatment process was carried out with varying different parameters such as weight ratios of aqueous ammonia to avicel cellulose, pretreatment duration and temperature. The amorphosity and crystallinity of the pretreated samples were then analyzed using XRD method (2θ in the range of 5°-80°). The XRD images showed that the highest decrystallization of cellulose was achieved at ammonium thiocyante, ammonia, avicel cellulose and water ratio of 1:4:1:1. Enzymatic hydrolysis was carried out using CTec-2 enzyme with varying hydrolysis time and enzyme loading. The yield of reduced sugar was analyzed using UV-Vis spectrophotometry. It was found that higher sugar yield was achieved using combined pretreated samples than untreated samples or ammonia pretreated samples alone. The highest sugar yield was found to be 35 mg/mL at 24 h hydrolysis time with enzyme loading at 5 mg protein/g glucan for the combined pretreated sample.

195c TEA and LCA of Renewable Hydrocarbon Fuels and Co-Products from Lignocellulosic Biomass Via 2,3 Bdo Fermentation, Separation, Dehydration and Oligomerization.

Marco Avendano, Qiang Fu, Jianpei Lao, Oindrila Gupta, Mi Lu, Sankar Nair, Matthew Realff

TEA and LCA analysis of Renewable Hydrocarbon Fuels and Co-products from Lignocellulosic Biomass via 2,3 BDO fermentation, separation, dehydration and oligomerization.

Production of fuels and chemicals from renewable feedstock is a well-known concept and after decades of commercial operation, first generation biorefineries have reached a high maturity level [1]. These first generation biofuels compete with food resources and have detrimental effects associated with land use change. As a result, second generation biofuels obtained from non-edible resources, such as agricultural residues, and other lignocellulosic sources have emerged as a promising alternative [2]. One example is the biochemical conversion of lignocellulosic biomass to hydrocarbon fuels. The National Renewable Energy Laboratory (NREL) performed a detailed techno-economic analysis (TEA) in 2018 of a pathway that produces a blend of diesel and naphtha from corn stover [3]. In this process 2,3 butanediol (BDO) is the key intermediate obtained via fermentation and adipic acid produced from lignin is the main co-product.

In our work we assessed the feasibility of this integrated biorefinery through an update of NREL's 2018 TEA and by performing a well-to-wheel (WTW) life cycle assessment (LCA). More specifically, we examined the effect of 1) adding a downstream separation system that concentrates BDO from the dilute fermentation broth and 2) varying fermentation conversion and BDO titer concentration. While NREL already assessed feasibility through a TEA, no LCA has been developed for this pathway. Furthermore, the flowsheet in place does not include any type of BDO separation step and to the best of our knowledge there isn't an environmental sensitivity study that accounts for varying fermentation conditions.

The suggested separation scheme concentrates BDO from a dilute fermentation broth and consists of two units: 1) a zeolitic adsorption system and 2) a pervaporation membrane. Experimental runs at a lab bench scale have been conducted for adsorption and pervaporation by collaborators at the Georgia Tech and Oak Ridge National Laboratory, respectively [2]. The experimental data has been used to parametrize these separation systems and implement them in the biorefinery process flowsheet. ASPEN Plus ® and Pyomo were used as the modelling tools.

An LCA of the overall process has also been developed, with global warming as the main impact category. The proposed LCA evaluates every stage in the production of 1GGE of fuel. This LCA accounts for the harvesting, handling, preprocessing and transport of the biomass, the process conversion inside the biorefinery and the end-use of the fuel at the vehicle's engine. In addition, we evaluated the effect of varying fermentation conditions in the overall LCA. We used NREL's 2018 report and our internal models for the mass and energy balances and for the remaining lifecycle inventory (LCI) process, to perform the other lifecycle impact calculations we used openLCA and Argonne's GREET model.

Our results show that the proposed separation system can concentrate BDO from 10wt% to 95wt%. This dramatically reduces the stream flowrate and causes a significant decrease in capital costs of downstream equipment and operating costs associated with heating and pressurizing. An update to the TEA indicates that the adsorption system reduces the MFSP of the fuel to $2.37/GGE. Moreover, the

LCA results indicate that for a base BDO titer of 100g/L, the WTW emissions of the fuel are 7.2kg/GGE, 40% lower compared to fossil fuels. Our results also indicate that emissions linearly decrease with increasing BDO titer but increase with conversion (at higher conversion we require less biomass, thus we obtain less lignin, which is utilized for internal utility consumption). Finally, the effect of adipic acid production through co-product displacement was also evaluated in the LCA.

References

[1] McKendry, P. (2002). Energy production from biomass (part 1): overview of biomass. Bioresource Technology, 83(1), 37–46.

[2] Segovia-Hernández, J. G., andamp; Sánchez-Ramírez, E. (2022). Current Status and Future Trends of Computer-Aided Process Design, Applied to Purification of Liquid Biofuels, Using Process Intensification: A Short Review. Chemical Engineering and Processing-Process Intensification, 108804.

[3] NREL, 2018. Ryan Davis, Nicholas Grundl, Ling Tao, Mary J. Biddy, Eric C. D. Tan, Gregg T. Beckham, David Humbird, David N. Thompson, and Mohammad S. Roni. Biochemical Deconstruction and Conversion of Biomass to Fuels and Products via Integrated Biorefinery Pathways. National Renewable Energy Lab.

[4] Church, A. L., Sun, N., Yan. J., Karp, E., Tan, E., McNamara, I., Freeman, C., Liu, J., Glezakou, V., Zhang, D., Dunn, J., Valentino, L. (2021). 2,3-Butandiol (BDO) Separations. Separations Consortium Technology Area Session: Performance-Advantaged Bioproducts, Bioprocessing, Separations, and Plastics.https://www.energy.gov/sites/default/files/2021-04/beto-18-peer-review-2021-bioproduct-church.pdf

195d Advanced Biorefinery Feedstocks from Waste Biomass: Technoeconomic Assessment of Integrated Air Classification-Hydrothermal Carbonization-Pelletization Processes

Md Tahmid Islam, Nepu Saha, Jordan Klinger, Toufiq Reza

Corn stover is one of the most abundant agronomic resources in the U.S., however, those are either incinerated or unprocessed in the field to avoid high logistical cost, economic forfeiture, and environmental complications. Air classification has been proposed to separate clean corn stover from waste corn stover. Waste corn stover can be hydrothermally carbonized (HTC) into solid binder called hydrochar. Clean corn stover from air classification can be pelletized with hydrochar in a high moisture pelleting process (HMPP). The final product of the integrated process, pelletized corn stover, is a conversion-ready biorefinery feedstock. The aim of this study is to investigate economic feasibility of integrated air classification-HTC-HMPP process. Different scenarios were considered for technoeconomic analysis to convert 100 tonne/day corn stover into HMPP pellets. While, optimized air classification and HMPPP process parameters are obtained from our prior studies, HTC process was optimized using two cases. Case I assumed a batch HTC process and case II assumed a continuous HTC process in the integration system. Results show that both cases are economically feasible, where case I reached its break-even years earlier than case II. Additionally, despite the greater capital investment requirement for case I, there was also a greater return on investment (ROI) compared to case II. A sensitivity analysis was also performed which revealed that the effect of solid content in air classified waste corn stover or reactor's feed, feedstock pumping cost, and total equipment cost were the influential parameters on breakeven selling price of the hydrochar.

Keywords: Corn stover; Hydrothermal Carbonization; Hydrochar; Economic Analysis; Sensitivity analysis;

195f Oxidative Fast Pyrolysis of Groundnut Shell (GNS) in the Catalytic and Non-Catalytic Mode for Biofuel Production

Geeta Kumari, Bhavin Soni

A fluidized bed fast pyrolysis system (8 kg/hr) was designed and developed at Sardar Patel Renewable Energy Research Institute (SPRERI). Groundnut shell (GNS) is selected as a biomass feedstock and Fe_2O_3, CaO, MgO and TiO_2 are selected as a catalyst for the fast pyrolysis experiment. The parameters for the trials were determined at 8 kg/hr feeding rate at the temperature range of 550-650 °C with an equivalence ratio of 0.1-0.2. The bio-oil yield with the existing parameter was observed at ~ 37 wt. % in non-catalytic mode. In catalytic mode, the bio-oil yield was observed at 31.34 wt. % with Fe_2O_3, 29.90 wt. % with CaO, ~27.23 wt. % with MgO and 28.30 wt. % with TiO_2. The organic phase in the bio-oil in catalytic mode is improved by 5-7 wt. %. The physical properties such as pH, solid content and calorific value of the bio-oil were evaluated. The pH of the bio-oil in the non-catalytic mode (4.05) is lower than in the catalytic mode (4.86 - 5.75) with all the catalysts, which indicates the catalytic bio-oil is less acidic than the non-catalytic bio-oil. The solid content in the bio-oil with the catalyst (3.98 – 4.13 wt. %) is less than the non-catalytic (5.57 wt. %) mode. The calorific value of the bio-oil was observed higher in catalytic mode (29.95 - 31.41 MJ/kg) in comparison to non-catalytic (29.89 MJ/kg) mode. The ^1H NMR analysis of bio-oil from sawdust and GNS reveals the presence of alkyl (methyl and methylene), alkenyl, aryl, allylic, alcoholic, and phenolic protons. The FTIR spectrum shows that different classes of organic compounds such as phenols, aliphatic alcohols, alkanes, alkenes, aromatics, ketones, aldehydes, nitrogen-containing compounds, carboxylic acids and halides are present in bio-oil.

195g TEA/LCA of Integrated Biochemical and Hydrothermal Processing of Corn Stover for Fuels and High Value Products

Bharathkiran Maddipudi, Khang Huynh, Vinod S. Amar, Katelyn Shell, Anuj Thakkar, Anuradha Shende, Sergio Hernandez, John E. Aston, Sandeep Kumar, Ram Gupta, Rajesh Shende

Globally, corn crop production is more than 1 bty, which generates significant amount of corn stover that comprised of leaves, stalks, and cobs of corn. An integrated processing approach comprised of biochemical and hydrothermal technologies is developed to minimize waste production and achieve production of both solid and liquid side products, which the viewpoint of offsetting the fuel cost. Corn bales were processed at the Biomass Feedstock National User Facility (BFNUF) and the obtained pulverized corn stover (low and high ash content) were pretreated to obtain unhydrolyzed solids (UHS). Hydrothermal liquefaction (HTL) of UHS was performed to achieve production of hydrochar, phenols and lactic acid. Hydrochar was further valorized into porous carbon, which was used for the energy storage application. Process engineering calculation and scale-up was performed for 1 tpd feedstock throughput. Technoeconomic analysis (TEA) and life-cycle analysis (LCA) were performed considering optimized integrated processing. Production rate of high value-added products such as char, phenols and carboxylic acids was included in the process economics calculations. ASPEN modeling was performed to scale-up the process to achieve fuel cost <$2/gge.

64a Evaluation of Combustion Properties of Corn Stover-Derived Solid and Liquid Biofuels By Cone Calorimeter

Md Tahmid Islam, Jordan Klinger, Toufiq Reza

This study investigated the combustion properties hydrothermally treated corn stover-derived solid and liquid biofuels using cone calorimeter. Hydrothermal treatments were performed at 200-350 °C for 30 min, where solid hydrochar was produced at lower treatment temperature and liquid biofuel was produced at higher treatment temperature. In order to test their combustion properties, six different heat fluxes (5-50 kW/m^2) were applied to biofuels to determine critical heat flux (CHF), ignition temperature, thermal response, peak heat release rate (pHRR), thermal hazard risk, and combustion efficiency. Results show that increasing the hydrothermal treatment temperature decreased the CHF and ignition temperature but increased the thermal resistivity. Increasing the heat flux decreased the thermal resistivity and ignition time of all the biofuels. Thermal hazard analysis indicated that the biofuels posed lower thermal risk than raw corn stover. Overall, hydrothermal treatment forms biofuel which are less flammable, more stable, and a lower thermally risk material for piloted ignition than raw corn stover.

Keywords: Corn stover; Biofuel; Cone calorimeter; Critical heat flux; Peak heat release rate;

64c Production of Sustainable Aviation Fuels from Lignocellulosic Biomass Feedstocks

Foster Agblevor, Hossein Jahromi

Decarbonization is a major thrust area worldwide to reduce carbon emissions and therefore reduce potential global warming. Aviation is a very important transportation industry in the world which faces decarbonization challenge. Unlike road transportation which is being electrified, electrification of air transport is near impossible. The alternative to decarbonizing air transport is through the production and utilization of low carbon fuels. Current aviation fuels are produced mostly from petroleum resources and a small fraction from biobased resources such as plant lipids and algae lipids which are not sustainable. Efforts to produce aviation fuels from lignocellulosic biomass is often very challenging because fuels from lignocellulosic biomass pyrolysis oils are very rich in naphthenes and low in long chain aliphatic hydrocarbons which are major components of the petroleum derived aviation fuels. We have investigated the catalytic pyrolysis and hydrodeoxygenation of the pyrolysis oils to produce long chain hydrocarbons in the range suitable for aviation fuels. We have achieved the production of the long chain aliphatic hydrocarbons from biomass pyrolysis oils through alkylation and opening of the aromatic carbon rings. Carbonyl compounds were used to alkylate aromatic rings and then opened them into long chain aliphatic hydrocarbons. These reactions were catalyzed by nickel on red mud catalysts in aqueous medium. Thus, catalytic pyrolysis oils produced from pinyon juniper wood were successfully converted into aliphatic hydrocarbons and alkylated aromatic, and naphthenes suitable for application as aviation fuels. Carbon chain length ranged from C_6 to C_{17} and both aqueous phase and organic phase pyrolysis oils were successfully used for the hydrocarbon production.

64e Evaluation of Ultra-High Temperature Resistant Preformed Particle Gels for Geothermal Preferential Fluid Flow Control

Tao Song, Thomas Schuman, Baojun Bai

In recent years, to reduce carbon emissions, more and more attention has been focused on utilizing indigenous, renewable energies, such as solar, tidal and geothermal energy. Geothermal energy as a cost-effective green energy has been deployed to provide electricity worldwide for more than 100 years. The principle of generating geothermal energy is to inject cold water into the deep subsurface geothermal reservoir through the injection well, where temperatures are high enough for power generation (150~200 °C), and the injected cold water is recovered as steam or hot water through the production well. The extracted heat can be used for power generation. However, natural or unwanted fractures generated during drilling and hydraulic fracturing could cause fast and excessive cold water production and reduce the heat exchanging time between the cold water and the geothermal reservoirs, which significantly deteriorates the heat extraction efficiency. Currently, there are no hydrogel products available to meet the harsh conditions of geothermal reservoirs. This work systemically evaluates a novel high-temperature resistant preformed water-swellable particle gel system (HT-PPG) for controlling the fluid and heat flow through fracture networks in the geothermal reservoirs. We assessed the effect of variables like temperature, pH, and salinity on swelling behavior, gel strength, long-term thermal stability and plugging efficiency of this novel HT-PPG. The dried gel particle can swell to over 20 times its original volume, and the fully swelled particle gel's elastic modulus can reach over 800 Pa. Additionally, the HT-PPG showed excellent long-term hydrolytic thermal stability. No syneresis was observed during the three months exposure at 175 °C. HT-PPG described in this work is a promising product for controlling the fluid and heat flow in fractures containing geothermal reservoirs.

64f Roles of Urea Inclusion Fractionation of Biodiesel on the Cold Flow Properties and Establishment of Feedstocks for Biochemicals and Biomaterials

Junli Liu, Bernard Tao

Biodiesel mainly consists of methyl esters derivatives of long-chain fatty acids and is viewed as the renewable alternative to petroleum diesel for carbon sequestration and mitigating environmental problems. However, its utilization is constrained by the poor cold flow properties. Cloud point, the onset temperature of thermal crystallization, can appropriately evaluate cold flow properties. The saturated components with high melting points significantly influence cloud points, and removing saturated components can effectively reduce the cloud point. Distillation, solvent thermal extraction, and thermal crystallization are the fractionation based on the phase equilibrium and own the disadvantages of low yields and slow production rate. A particular method was performed to fractionate FAME at room range temperatures by forming solid urea inclusion compounds (UIC). Urea inclusion fractionation provides a way to reduce the cloud point of mixtures of FAME by preferentially removing high melting-point linear saturated FAME components. Since soybean oil is the primary feedstock for biodiesel in the United States, the effects of the mass ratio of urea to FAMEs, the mass ratio of methanol to FAMEs, and the type of solvents on urea inclusion fractionation were carried out on FAMEs from soybean oil in urea inclusion fractionation. In addition, FAMEs from other sources were used to test the feasibility of urea inclusion fractionation on the cold flow properties. According to the experiments, the cloud point of biodiesel from soybean oil after urea fractionation could be as low as -52 ^{o}C. After twice urea inclusion fractionation for palm oil biodiesel, the cloud point can reach as low as -17 ^{o}C. In addition, the effect of urea inclusion on the biodiesel was also evaluated in the cetane number, oxidation stability, nitrogen oxides emission, etc. The separation efficiency and selectivity were measured according to the driving force in urea inclusion fractionation. Compared to distillation solvent thermal extraction and thermal fractionation, urea inclusion fractionation could reach high yields and good cold flow properties.

64g Novel Supercritical Biodiesel Plant Design and Process Scale-up

Shyam Paudel, Caleb Moellenhoff, Haider Al-Rubaye, Joseph Smith

Current energy prices, national energy security, and global climate change continue our efforts to find efficient advanced fuels choices. Traditional biodiesel processes are Acid and base-catalyzed processes, they use traditional continuous stirred-tank type reactors, and have distillation schemes to separate material constituents. In past years, we tested a lab-scale no-catalyst supercritical biodiesel process using waste cooking oil and methanol as feed stock. Recently, we identified a noble separation mechanism to separate free fatty acids (FFA), FAME andamp; Glycerol, and glycerol andamp; methanol. We have also scaled up the continuous tubular reactor. The process plant has heat exchangers that are in networks and integrated into the plant loop.

A 1000 kg/day biodiesel plant will operate in a continuous operation and is made of eight modular nodes. These nodes are envisioned, designed, and simulated in Sketch up (3D design software). The process nodes are modular and easily transportable and connectable. Our vision is to transport in the back of the tractor-trailer. Mass and energy balances of major process units (separation, andamp; reaction) are simulated using Aspen Plus^{TM} software. Our next steps include a comprehensive HAZOP study, control plan, bill of material, a production-ready Pandamp;ID, process simulation with Aspen Plus, and plant in a 3D sketch.

453a Direct Methane Conversion By Atomically Thin Platinum Nanolayers on Two Dimensional Metal Carbides

Yue Wu

Efficient and direct conversion of methane to value-added products has been a long-term challenge in natural gas and shale gas applications. In this presentation, we show that atomically thin nanolayers of Pt with a single or double atomic layer thickness, supported on a two-dimensional molybdenum titanium carbide (MXene), catalyse non-oxidative coupling of methane to ethane/ethylene (C2). Kinetic and theoretical studies, combined with in-situ spectroscopic and microscopic characterizations, demonstrate that Pt nanolayers anchored at the hexagonal close-packed sites of the MXene support can activate the first C–H bond of methane to form methyl radicals that favour desorption over further dehydrogenation and thus suppress coke deposition. At 750 °C and 7% methane conversion, the catalyst runs for 72 hours of continuous operation without deactivation and exhibits >98% selectivity towards C2 products, with a turnover frequency of 0.2–0.6 s^{-1}. Our findings provide insights into the design of highly active and stable catalysts for methane activation and create a platform for developing atomically thin supported metal catalysts.

453b Oxidative Coupling of Methane Catalysts: Developing Structure-Property Relationships for High-Performance Metal Oxide Systems

Mariano D. Susman, Matteo Ferri, Raffaele Cheula, Hien N. Pham, Abhaya K. Datye, Sivadinarayana Chinta, David West, Matteo Maestri, Jeffrey Rimer

It was fourty years ago that Keller and Bhasin[1] first reported the oxidative coupling of methane (OCM) in their seminal work. They envisioned the oxidative conversion of methane present in abundant gas resources (e.g., shale and natural gas) into reactive chemical commodities such as ethane, ethylene, propane, etc. (C_{2+} hydrocarbons). Despite significant efforts over the years, commercial implementation of OCM has been restrained, most significantly due to limitations in finding a catalyst leading to a sufficiently high performance as well as a suitable reaction process amenable to the high temperature conditions needed (>700 °C). In spite of a myriad of catalysts that have been studied, the reaction mechanism largely remains elusive, including the nature of active sites, and important molecular reaction steps as are the oxygen and methane activation steps.

In this presentation, we will discuss a collaborative effort in catalyst design, testing, and computational modeling to explore two main research strategies: (1) developing structure-catalytic performance relationships using simple metal oxides, and (2) the evaluation of mixed metal oxide catalyst performance under autothermal reaction conditions. We have developed methods to prepare metal oxide catalysts with various crystal morphologies, and used these materials to assess OCM performance. The results of these studies were rationalized by computational modelling, thus leading to the conclusion that metal oxide crystal habit plays a significant role in OCM catalysis. Here, we will discuss these combined efforts and show how optimized metal oxide systems utilized in an autothermal reactor design enable high-performance OCM under conditions that are close to being commercially viable.

References:

[1] Keller, G. E.; Bhasin, M. M. Synthesis of Ethylene via Oxidative Coupling of Methane. I. Determination of Active Catalysts. *J. Catal.* **1982**, *73*, 9– 19.

453c Pt-Based Monolithic Catalysts for Oxidative Coupling of Methane: Effect of Catalyst Formulation and Operation Parameters

Sven Schardt, Ahmet Celik, Jaspreet Chawla, Simon Bastian, Stephan A. Schunk, Patrick Lott, Olaf Deutschmann

Introduction

Over the decades, significant research has been conducted on methane as possible platform chemical for various reactions [1,2]. Amongst these, the oxidative coupling of methane (OCM) has been subject to many studies elucidating the influence of the active material (e.g. $MnNa_2WO_4$, Pt, and other metals) on the formation of C_2 species [3-5]. Recently, Chawla *et al.* [4] investigated a monolithic Pt/Al_2O_3 catalyst for acetylene formation at high temperatures and used computational methods to understand the complex OCM mechanism over Pt, which can ultimately contribute to improving the catalyst's activity and product selectivity. Furthermore, studies on 3D printed monoliths by Bogdan et al. [5] suggest a strong correlation between catalyst geometry and C_2 yields.

In our present study we investigate the influence of the monolithic structure together with the composition of the washcoat to reach higher acetylene selectivities. Changing the monolith length and the monolithic channel diameter as well as the support material and the washcoat thickness showed pathways to achieve higher acetylene selectivities.

Materials and Methods

The Pt catalysts were prepared by incipient wetness impregnation of the calcined metal oxide support materials Al_2O_3 and ZrO_2. The received powder was coated onto a cordierite honeycomb monolith and fixed in a quartz glass tubular reactor. A heat shield in front of the catalyst ensured efficient heat transfer. The reactor was placed in a furnace and fed with a gas mixture consisting of methane, oxygen, and nitrogen to ignite the autothermal reaction at 500 °C. The exhaust gas was monitored by means of FTIR. Optimal reaction parameters were found by varying the N_2 dilution between 85% and 30% and CH_4:O_2 ratios of 1.1 to 2.4 at a GHSV range of 68,000 h^{-1} to 1,200,000 h^{-1}. In addition, the effect of different monolith cell densities (300, 400 and 600 cpsi) and varying washcoat composition was determined.

Results and Discussion

Figure 1 shows the influence of the N_2 dilution on the OCM over a Pt/ZrO_2 catalyst. The data suggest a strong connection between the N_2 content in the feed and the downstream temperature as well as the C_2 selectivities. With rising share of reactive gases (CH_4 and O_2), higher temperatures are achieved that lead to a higher C_2 species formation. However, a maximum at 55 vol.-% and 40 vol.-% of N_2 for ethylene and acetylene selectivity, respectively, is visible, whereas the highest temperature (~1180 °C) is found at 35 vol-%, suggesting temperature optimum for the gas phase reactions. Table 1 shows the influence of washcoat thickness over Al_2O_3 supported catalysts. As all listed catalysts exhibit a Pt loading of approx. 30 g ft^{-3} and approx. identical particle size, higher Pt content in the used powder leads to a thinner washcoat, whereas the washcoat is approx. half the size for doubled Pt content (washcoat thickness: 0.5 wt.-% > 1 wt.-% > 2 wt.-%). The best performance was observed for the 1 wt.-% Pt/Al_2O_3 catalyst; the 2 wt.-% sample leads to a comparable total C_2 selectivity but less acetylene in the product stream. Using a 0.5 wt.-% Pt/Al_2O_3 coating, exhibiting the thickest washcoat of the investigated catalysts, leads to more (partial) oxidation of methane and less C_2 formation. Additionally, Table 1 shows the effect of different monolith geometries, proving 400 cpsi monoliths as most beneficial for OCM over Pt. By using smaller or bigger channel diameters ethylene formation seems promoted, but the overall desired C_2 product formation is lowered. Looking at all cases depicted in Table 1, the influence of the reaction temperature (implied by $T_{Downstream}$, measured 5 mm behind the catalyst) becomes visible. Although the highest temperature leads to the highest acetylene selectivity (1 wt.-% Pt; 400 cpsi), it does not seem to be the only important parameter. When comparing 0.5 wt.-% Pt; 400 cpsi with 1 wt.-%; 300 cpsi, a higher temperature with higher CO selectivity is observed for the first sample versus a lower temperature with a higher C_2 selectivity for the latter one, respectively. This leads to the conclusion, that the monolith's structure, including the washcoat's composition, has a crucial impact on the product formation of the Pt-OCM.

Taking all the investigated parameters into consideration the 1 wt-% Pt/ZrO_2 catalyst on 400 cpsi exhibited the best results with 11.2% C_2 selectivity and 89.1% CH_4 conversion at a space velocity of 306,000 h^{-1} and a CH_4:O_2 ratio of 1.1.

References

[1] R. Franz, E.A. Uslamin, E.A. Pidko, *Mendeleev Commun.*, **2021**, 31, 584-592.

[2] O. Deutschmann, L.D. Schmidt., *AIChE J.*, **1998**, 44, 2465.

[3] G.E. Keller, M.M. Bhasin, *J. Catal.*, **1982**, 73, 9-19.

[4] J. Chawla, S. Schardt, S. Angeli, P. Lott, S. Tischer, L. Maier, O. Deutschmann, *Catalysts*, **2022**, 12 (2), 189.

[5] E. Bogdan, B. Michorczyk, A. Rokicińska, M. Basta, M. Myradova, P. Kuśtrowski, P. Michorczyk, *Appl. Surf. Sci.*, **2021**, 553, 149554.

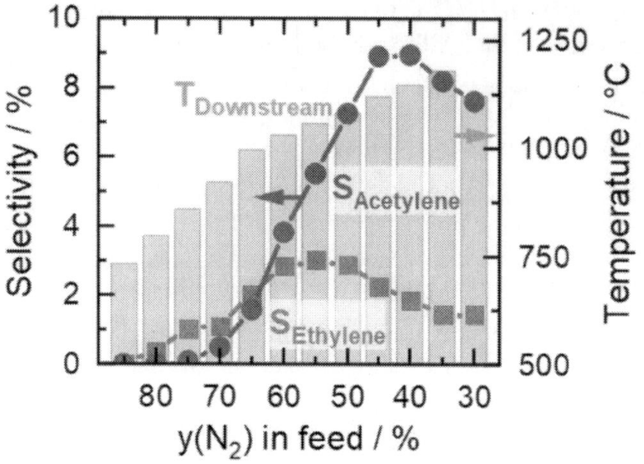

Figure 1: C_2 selectivities and downstream temperature in dependency of the N_2 content in feed for a 1 wt.-% Pt/ZrO_2 catalyst on a 400 cpsi cordierite monolith at a GHSV of 306,000 h^{-1} and a $CH_4:O_2$ ratio of 1.1.

Table 1: Selectivities for acetylene, ethylene, and CO as well as methane conversion and downstream temperature for different Pt/Al_2O_3 systems with a Pt loading of 30 g ft^{-3} at a GHSV of 460,000 h^{-1}, a $CH_4:O_2$ ratio of 1.1, and a N_2 dilution of 49 vol.-%. Selectivities are based on methane. Reference of 1-wt.% Pt/Al_2O_3 and 400 cpsi is shown twice for comparability.

Al_2O_3 catalysts	$S(C_2H_2)$ / %	$S(C_2H_4)$ / %	S(CO) / %	$X(CH_4)$ / %	$T_{Downstream}$ / °C
0.5-wt.% Pt; 400 cpsi	6.7	1.2	69.3	91.4	1141
1-wt.% Pt; 400 cpsi	9.0	1.5	65.2	90.6	1207
2-wt.% Pt; 400 cpsi	8.3	2.2	64.3	89.0	1105
1-wt.% Pt; 300 cpsi	7.2	1.8	56.6	90.3	1094
1-wt.% Pt; 400 cpsi	9.0	1.5	65.2	90.6	1207
1-wt.% Pt; 600 cpsi	6.3	2.7	58.3	84.7	1064

453d New Insights into the Effect of CO2 on Oxidative Coupling of Methane (OCM)

Yonggang Cheng, Pedro Mendes, Joris Thybaut

CO_2 being a soft oxidant and one of the major side products in oxidative coupling of methane (OCM), its impact on OCM is of great interest. Up to now, significant efforts and progresses have been made in this respect, however, no consensus has yet been reached. On the contrary, the effect of CO_2 on OCM appears to be diverse and catalyst-dependent[1,2]. In this work, new insights into the effect of CO_2 on OCM have been acquired based on experimental investigations.

The reaction performances exhibited by two benchmark OCM catalysts, i.e. La-Sr/CaO and NaMnW/SiO_2, were investigated in both O_2-rich and O_2-lean conditions. Specifically, in O_2-rich conditions CO_2 was added by substituting inert balance gas while in O_2-lean conditions O_2 was substituted by equal amount of CO_2. In addition, a gas mixture with equimolar amount of C_2H_6 and CO_2 was also fed in the attempt to study the occurrence of CO_2 assisted oxidative dehydrogenation of ethane (CO_2-ODH) on the OCM catalysts.

The catalyst-testing results in O_2-rich conditions are shown in figure 1. Overall, CO_2 co-feeding has a detrimental impact on both catalysts in terms of C_2 selectivity. This effect is more pronounced on La-Sr/CaO, which is deactivated already by 8% CO_2. In contrast, NaMnW/SiO_2 has a higher tolerance to CO_2 up to 20%. Contrary to the observations in O_2-rich conditions, a positive effect of CO_2 has been observed in O_2-lean conditions. Compared with the reference test without co-feeding CO_2, the selectivity towards C_2H_4 increases, and this effect becomes more significant with the CO_2 fraction. The increase in C_2H_4 selectivity is always achieved at the expense of C_2H_6 selectivity, and is accompanied by a decrease in CO_2 selectivity. Seemingly, the addition of CO_2 in O_2-lean conditions promotes the dehydrogenation of C_2H_6 to C_2H_4. The results obtained by feeding a CO_2-C_2H_6 mixture are summarized in table 1. In comparison with the blank tests, the conversion of C_2H_6 and CO_2 are evidently increased in the presence of a catalyst, verifying the occurrence of CO_2-ODH on the OCM catalysts.

Based on the experimental investigations, a mechanism of the CO_2 effect on OCM is proposed. Firstly, CO_2 as a soft oxidant competes for the active sites with O_2 and produces a new type of oxygen surface species [O*]'. Different from the oxygen surface species [O*] generated by O_2 chemisorption, [O*]' is less active but is capable of triggering CO_2-ODH. In O_2-rich conditions, the competition between O_2 and CO_2 makes O_2 less consumed in catalytic reactions, which lowers the catalyst activity and indirectly intensifies the gas-phase oxidation of C_2 species eventually leading to the decrease of C_2 yield. Therefore in this condition, CO_2 only exhibits negative effect. On the other hand, in O_2-lean conditions, the low O_2 fraction makes it possible to have OCM and CO_2-ODH taking place in parallel. The [O*]' promotes the conversion of C_2H_6 to C_2H_4 and prevents C_2H_6 from being further oxidized into CO_2. In the meantime, with lower O_2 fraction, the C_2 species also have higher chance to survive from the gas-phase oxidation. Consequently, positive effect of CO_2 can be observed in the O_2-lean condition.

With the investigations conducted in this work, it is concluded that the effect of CO_2 is not only catalyst-dependent but also conditions-dependent.

Reference

1. Yide, X., Lin, Y. andamp; Xiexian, G. Effect of basicity and adding CO_2 in the feed on the oxidative coupling of methane over K_2O and SrO promoted La_2O_3/ZnO catalysts. *Appl. Catal. A Gen.* **164**, 47–57 (1997).
2. Shi, J., Yao, L. andamp; Hu, C. Effect of CO_2 on the structural variation of Na_2WO_4/Mn/SiO_2 catalyst for oxidative coupling of methane to ethylene. *J. Energy Chem.* **24**, 394–400 (2015).

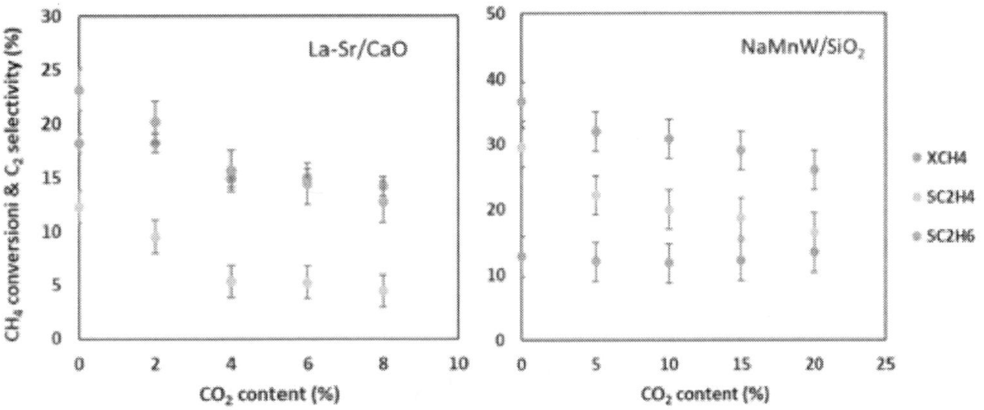

Figure 1: Catalyst testing results in O_2-rich conditions (Conditions: T = 800 °C, CH_4/O_2 = 4, P = 1 bar, ~ 16% X_{CH4} for La-Sr/CaO, ~ 12% X_{CH4} for NaMnW/SiO₂)

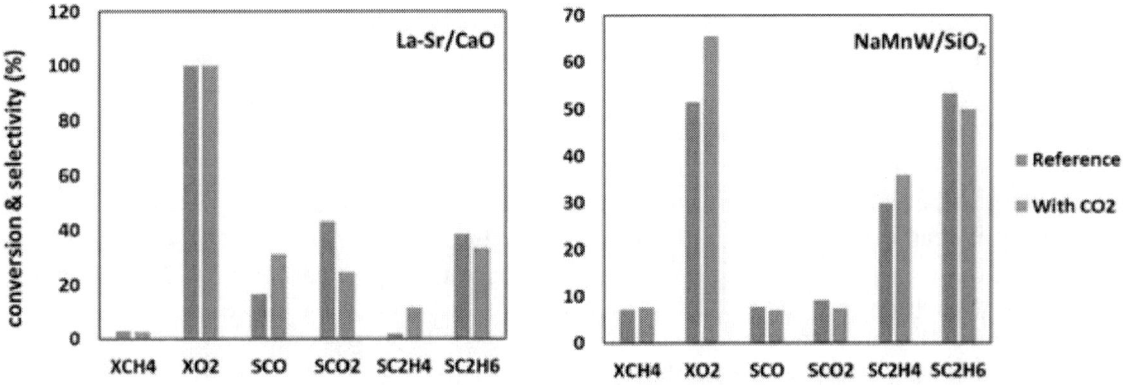

Figure 2: Catalyst testing results in O_2-lean conditions (Conditions: T = 800 °C, P = 1 bar, O_2/CO_2 = 0.5%/9.5% for La-Sr/CaO, , O_2/CO_2 = 1.0%/9.0% for NaMnW/SiO_2)

Table 1: The testing results with gas mixture of CO_2 and C_2H_6

	X_{CH4}	X_{CO2}	S_{C2H4}[c]	S_{CO}[c]
Blank 1[a]	29.7%	1.9%	99.7%	0.3%
La-Sr/CaO[a]	36.7%	10.6%	99.5%	0.5%
Blank 2[b]	24.3%	1.3%	99.4%	0.6%
NaMnW/SiO_2[b]	27.3%	4.6%	99.2%	0.8%

a: the volumetric flow rate is 142 ml/min

b: the volumetric flow rate is 160 ml/min

c: the selectivity in terms of C_2H_6, which is calculated as:

$$S_{C_2H_4} = \frac{2 \cdot F_{C_2H_4}}{(2 \cdot F_{C_2H_4} + F_{CO} - (F_{CO_2,out} - F_{CO_2,in}))}$$

$$S_{CO} = \frac{F_{CO}}{(2 \cdot F_{C_2H_4} + F_{CO} - (F_{CO_2,out} - F_{CO_2,in}))}$$

453e Non-Catalytic Direct Partial Oxidation of Methane to Methanol in a Microreactor

Kelly Cohen, Justin Blanchard, James Dorman, Kerry Dooley

This project uses an integrated mixer-reactor-heat exchanger microtube scalable module for the non-catalytic direct partial oxidation of methane in natural gas (NG) to methanol at elevated pressures (70-95 bar), reaction temperatures of 380-460°C, reactant CH_4:air ratios of 1.5-2.9 and reaction zone residence times on the order of 1 min.

Currently, methane in remote locations is often flared in large quantities rather than transported or converted to the more easily transportable methanol. However, if this gaseous methane were converted to liquid methanol at the wellhead, it could be transported in ways other than pipeline, or used onsite. For example, methanol is added to pipeline gas at remote locations to prevent hydrate formation in pipelines, and is a component used in water purification packages.

The microtube reactor/heat exchanger system for this process consists of a methane/air mixing, preheating and reaction section and a quenching section. In the quenching section, the heat transfer is from tubes to shell. The shell-side flow rate is controlled to attain the desired reaction and final (~250°C) process fluid temperatures. The entire reactor, tubes and shell, are made of Inonel 625 alloy. Alloy I625 is a high Ni-Cr steel (60 wt% Ni, 22 wt% Cr). To passivate the metal and reduce rates of total oxidation, the tubes are carbide-coated in a process based on the decomposition of propane at ~700°C. Several other coating schemes were tested along with several other potential tube metals (e.g., sulfiding instead of carbiding, and also metals C1010 mild steel, 304L, 904L, Nichrome and pure Ni) in coating experiments using coupons. The reasons for the choice of I625 and the carbiding process will be discussed.

For the current microreactor system, methane conversions of 12.5% have been achieved with a methanol selectivity of 80% at 417°C, 80 bar, 0.82 min space time and a CH_4:air molar ratio of 2.9. However, at 0.50 min and molar ratio 1.5 the methanol selectivity is only 16% (CH_4 conversion 14.5%, 417°C, 83 bar). So while methanol can be made selectively in the microtube reactor system, its selectivity can vary over a wide range depending upon certain reaction conditions.

453f Electrochemical Oxidation of Methane to Methanol on Electrodeposited Transition Metal Oxides Under Ambient Temperature

Kangze Shen, Yu-Chao Huang, Joonbaek Jang, Carlos Morales-Guio

Methane is the primary constituent of natural gas and broadly utilized as an essential feedstock and fuel in chemical manufacturing units and energy transformations. From the current energy outlook, natural gas is the prominent option for fuel, energy, and feedstock in the chemical and petroleum industries. Methane in associated petroleum gas cannot be transported economically from remote oil fields and is thus usually flared resulting in significant greenhouse gas (GHG) emissions and energy losses leading to adverse influences on the climate, environment and energy markets. In this regard, electrochemical oxidation of methane to value-added chemicals such as methanol under ambient temperatures via modular systems is of great interest.

Research in the electrochemical oxidation of methane under ambient temperature using transition metal oxides as electrocatalysts has shown significant advances over the last decade. Binary transition metal oxides of NiO/ZrO_2 in a carbonate electrolyte have been shown to catalyze the electrochemical oxidation of methane. An analogous example utilizing chemical precipitation of Co_3O_4/ZrO_2 nanocomposite and Co_3O_4 powder/ZrO_2 nanotubes has been shown to produce higher alcohols such as 1-propanol and 2-propanol. Introducing zirconia by using co-precipitation to unary transition metal oxide as catalyst has shown to promote the methane oxidation in the presence of carbonate ions enabling the system to operate at room temperature. Intriguingly, the faradaic efficiencies (FEs) in some of these systems exceed 100% implying that chemical reactions between methane and stoichiometric oxidants could be responsible for the production of a large fraction of the oxygenates. Oxidation products could be also the result of the degradation of carbon conductive materials and binders added in the preparation of catalyst inks.

Methane oxidation reaction via electrochemistry is more favorable than water oxidation as it requires a less positive potential (E_0=0.58V vs SHE). However, because methane activation and the regeneration of oxidative species on the surface of the electrode are kinetically sluggish, this transformation may require higher overpotentials to drive the rate-limiting step at appreciable rates. At high overpotentials, the oxygen evolution reaction (OER, E_0=1.23V vs SHE) competes with the methane oxidation reaction and the high oxidation potential result in the over-oxidation of methane derived products to carbon dioxide and carbon monoxide which are more thermodynamically favorable. Accordingly, the major challenge is to develop electrocatalysts that can not only activate the C-H bond of methane at low overpotentials but also prevent the overoxidation of desirable products such as methanol. Many studies within the existing literature show inconsistent results due to the use of catalysts of varying morphology and composition as well as the use of different electrolyte compositions. This highlights the difficulties to determine product distribution accurately for methane partial oxidation and results in a lack of fundamental understandings of the kinetics associated with these electrochemical systems. The discussion of reaction mechanisms, theory-driven catalyst discovery, and optimization of reaction environments must be built upon reliable experimental evidence, which are still in short availability. In this study, a one-step electrodeposition method has been developed for the preparation of thin film transition metal oxides as model electrocatalysts without binders or carbon conductive materials. A systematic understanding of the electrochemistry in methane oxidation towards methanol is gained through kinetic studies for CH_4 partial oxidation in a customized gas-tight cell with a rotating cylinder electrode (RCE). Product analysis using NMR shows methanol and acetate are the major products during the oxidation. Additionally, by collecting and analyzing the liquid products with a 20-min interval, fluctuations in concentrations of products over time are observed thus highlighting the dynamic nature of partial oxidation studies. This is the case particularly for experiments where many of the liquid products such as methanol accumulate in the electrolyte and can be further transported back to the electrode surface to be oxidized. To further understand these phenomena, different factors including catalyst porosity, the applied potentials, the operating temperature, and the rotational speed of the electrode were systematically studied. Different unary transition metal oxides (CoO_x, NiO_x, CuO_x and MnO_x) were discovered to be active for methane partial oxidation with this simpler and cleaner fabrication method. Acetate is detected at low overpotentials near the thermodynamic minimum and methanol is the favored product at intermediate overpotentials, while high overpotentials result in the over-oxidation of the produced methanol and the production of oxygen. The temperature also plays an important role in both the activation of methane and the over-oxidation of the produced methanol. Furthermore, thicker catalyst films lead to more significant over-oxidation of methanol due to the highly porous structures and thus the longer paths for methanol to exit the catalyst layer. Higher electrode rotation speed of the RCE increase the cycling time of the concentration profiles in the electrochemical cell by affecting convection in the cell. These two last factors (catalyst porosity and electrolyte convection) are both correlated with the effect of mass transfer, implying the importance of understanding mass transport of the relevant species in partial oxidation studies of methane. To better illustrate the whole process, a numerical model describing the local concentrations of all species is built. From the evidence shown above, we propose that methane needs to go through a thermal activation step during the electrochemical oxidation and that the transport of methanol away from the electrode will be sufficient to obtain high selectivity to partial oxidation products. The experimental results as well as the numerical model in this study demonstrate a new approach tor catalyst fabrication and benchmarking and may inspire further studies of the electrochemical partial oxidation of methane under well defined conditions of mass, heat and charge transport.

453g Ethane Dehydrogenation Using Liquid Metal Catalysts

Aime Laurent Twizerimana, Kaylen Ocampo, Aaditya Hari Bharanidharan, Kayle McGaughy, Mohammad S. Masnadi, Goetz Veser

Ethylene is a major building block in the chemical industry, with an annual production of 172 million tons in 2021. Due to the availability of abundant shale gas, ethylene is produced in the U.S. via steam cracking of ethane, i.e. via mixing ethane with steam at high temperature (~ 800) where it is dehydrogenated to ethylene. Although this process is widely used commercially, it suffers from multiple issues, including coking due to carbon deposition on the reactor walls and high emissions due to the required heating to high reaction temperature. Formation of the coke layer on the reactor walls results not only in reduced ethylene yields but also in further insulation of the reactor from the (external) heating source, and hence a further increase in the required energy input to maintain a steady reaction temperature. Reducing the impact of coke formation could hence have major impact on the energy intensity of this process. We are studying the use of liquid metals as an alternative reaction medium which has the potential to address this issue.

Liquid metals, i.e. metals with a low melting point, have recently emerged as an interesting novel reaction medium. Liquid metals combine a high heat capacity with excellent resistance to coking by separating the formed coke from the catalytically active molten metal due to density differences. Liquid metals hence have the ability to resist deactivation due to coke formation and thus offer the potential for continuous operation as well as continuous removal of coke without disrupting the production process, without need for steam, and without the CO_X emissions associated with the decoking process.

Here, we present initial data evaluating bismuth-based liquid metal systems for ethane dehydrogenation to ethylene. Molten Bi baths were doped with small amounts of catalytically active metals. While pure Bi showed no detectable activity, doping with Ni and In induced catalytic activity. Addition of Ni favored the formation of methane and coke, in agreement with the well-known carbon scission activity of solid Ni catalysts. In contrast, doping with In favored the desired ethylene production. In our presentation we will present characterization of the liquid metal systems, preliminary reactor design considerations for these novel reaction systems, as well as results from the catalytic studies.

453h Ethane Oxidative Dehydrogenation By CO_2 over Stable Csru/CeO_2 Catalyst

Xiaoyan Wang, Yuxin Wang, Brandon Robinson, Jianli Hu

Accompanied by the development of modern industry, the demand for light olefins including ethylene, propylene, butene, and butadiene continues to increase. Light olefins are important intermediates for the production of polymers (polyethylene, polypropylene, etc.) and rubber (styrene butadiene rubber, polybutadiene rubber, etc.). Currently, the primary process to produce light olefins is based on steam cracking of petroleum liquids such as naphtha and distillate fuel oil. It is an energy-intensive process, requiring high operating temperatures (750-850°C) and pressure (1.0-4.5 MPa). As the feedstock becomes lighter, the product distribution is shifted to ethylene.

Besides petroleum liquids, ethane is a promising feedstock for ethylene production by steam cracking. Oxidative dehydrogenation (ODH) of ethane by CO_2 is an energy-efficient way of utilizing abundant shale gas while mitigating greenhouse gas impact.

In this study, oxidative and non-oxidative dehydrogenation (EDH) of ethane were investigated and compared over Cs-promoted Ru/CeO_2 catalyst. The presence of CO_2 improved the ethane conversion to ethylene and led to the formation of C_3 and C_4 olefins as well (Figure 1). The olefin selectivity and yields in ODH were affected by reaction variables including CO_2/C_2H_6 ratio, temperature, and feed rate. The CsRu/CeO_2 catalyst exhibited excellent stability in the ODH reaction. Mechanistically, the interaction between Cs and Ru with ceria in the crystalline structure facilitated reduction and oxidation cycles. It was postulated that in the oxidation cycle, Ce^{3+} had a strong tendency to react with CO_2 to form Ce^{4+}, causing C=O bond scission, whereas in the reduction cycle, Ce^{4+} was reduced by hydrogen during ethane dehydrogenation to form Ce^{3+}. The redox cycle can be repeated during the ODH reaction keeping the catalyst active. Furthermore, in the ODH reaction, the CsRu/CeO_2 catalyst exhibited resistance to carbon deposition, evidenced by the carbon balance measurement.

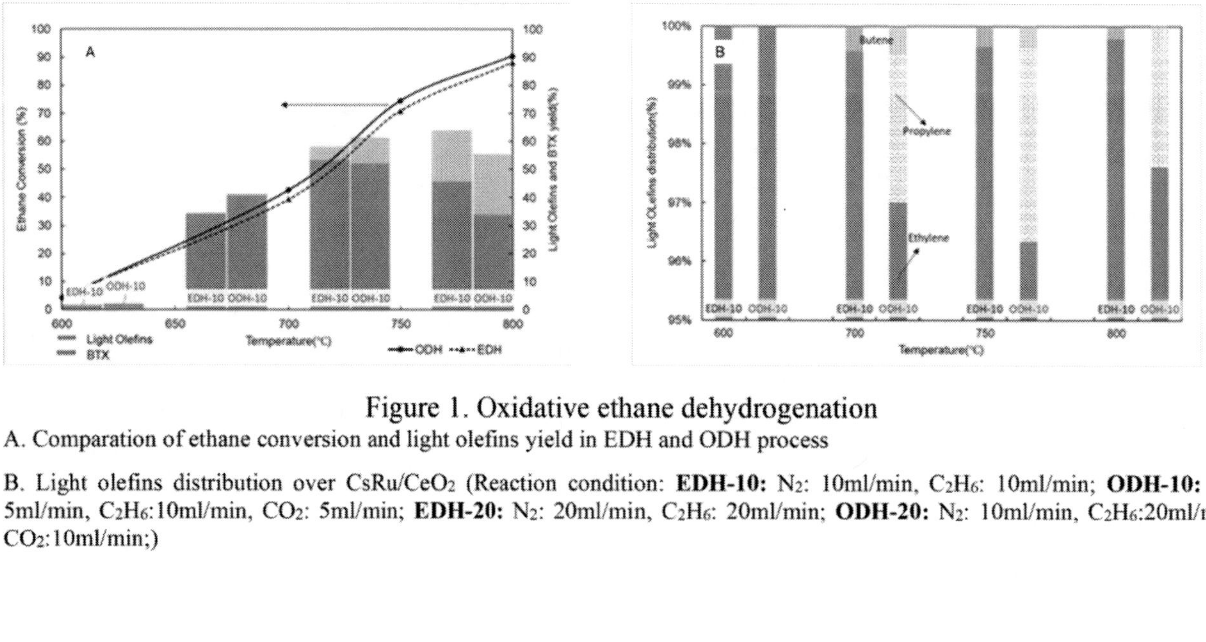

Figure 1. Oxidative ethane dehydrogenation

A. Comparation of ethane conversion and light olefins yield in EDH and ODH process

B. Light olefins distribution over CsRu/CeO_2 (Reaction condition: **EDH-10:** N_2: 10ml/min, C_2H_6: 10ml/min; **ODH-10:** N_2: 5ml/min, C_2H_6:10ml/min, CO_2: 5ml/min; **EDH-20:** N_2: 20ml/min, C_2H_6: 20ml/min; **ODH-20:** N_2: 10ml/min, C_2H_6:20ml/min, CO_2:10ml/min;)

AUTHOR INDEX

Abbas, A. S. 18
Abdelsayed, Victor 108
Abedin, Ashraf 108
Agblevor, Foster 88, 124
Ager III, Joel W. 34
Akbulut, Mustafa 89
Al Shalabi, Emad W. 86
Al-Abedi, Hasan 59
Al-Khedhair, N. 18
Almansa, Xavi Fonoll 7
Al-Musleh, Easa 69
Alomair, O. 18
Alomair, Osamah 95
Al-Rubaye, Haider 59, 127
Al-Shamlan, Yousef 27
Alshammri, Ahmed 95
Amar, Vinod S. 122
Angelopoulos, Anastasios 61
Anjum, Nooram 96
Asadi, Javad 81
Aston, John E. 122
Ates, Harun 71
Avendano, Marco 118
Bagemihl, Isabell 64
Bai, Baojun 125
Bai, Xinwei 106, 108
Balapitiya, Elizabeth 50
Balyan, Sonit 75, 114
Banerjee, Subhodeep 58
Barecka, Magda 34
Barton, David G. 12
Bartuel, Jean-Jacques 100
Bassi, Amarjeet 37
Bastian, Simon 130
Bayat, Hengameh 8
Belden, Elizabeth 5
Bell, Ian H. 99
Bharanidharan, Aaditya Hari 139
Bhat, Bhargavi 89
Bhatia, Divesh 28
Bhatraju, Chaitanya 64
Blanchard, Justin 136
Bloxham, Joseph 98
Bluck, David 100
Bongartz, Dominik 65
Bose, Monidipa 117
Brady, Casper O. 1
Brée, Luisa Carola 65
Breitkreuz, Christian F. 47
Brewer, Catherine 8
Broadbelt, Linda 72
Brown, Sean 110
Brushett, Fikile R. 66
Burger, Jakob 47
Burrows, Lauren 103
Caiola, Ashley 110, 114
Campestrini, Marco 100
Celik, Ahmet 130
Chanoi, Zachary 79
Chawla, Jaspreet 130
Chen, Chau-Chyun 96
Cheng, Feng 5
Cheng, Yonggang 133
Cheng, Zhuo 111
Cheula, Raffaele 129
Chiang, Yet-Ming 66
Chinta, Sivadinarayana 129
Clark, Michael B 12
Clingenpeel, Amy C. 102
Cohen, Kelly 136
Crandall, Dustin 103
Cronin, Dylan 7
Danghyan, Vardan 78
Dantu, Ryan 90
Datye, Abhaya K. 129
Dehghanizadeh, Mostafa 8
Depicker, Joost 12
Deutschmann, Olaf 130
Dhawan, Ashish 50
Ding, Dong 62
Ding, Hanping 62
Ding, Yanxu 85
Doan-Nguyen, Vicky 111
Dooley, Kerry 136
Dorman, James 136
Dreyer, Daniel 50
Dubey, Pranjul Mani 61
Dutta, Abhijit 2
Ebeling, Katharina Maria 65
Edmond, Kazem V. 102
Eichmann, Shannon 68, 71
Elsharkawy, Adel 95
Enick, Robert M. 103
Fan, Liang-Shih 74, 80, 111
Fang, Junchuan 61
Fantke, Peter 115

Ferre, Alvaro ... 47
Ferri, Matteo ... 129
Forner-Cuenca, Antoni ... 66
Fu, Qiang ... 118
Garcia, Freddy ... 100
Gausmann, Marcel ... 35
Giles II, Jefferson T. ... 102
Giles, Neil ... 98
Goertz, Jonas ... 35
Gondane, Ayush ... 93
Goodman, Angela ... 103
Gopinadhan, Manesh ... 102
Griffin, Michael B. ... 2
Grohol, Daniel ... 12
Gross, Adam S. ... 102
Guffey, Cassandra ... 98
Guo, Mond F. ... 1
Gupta, Oindrila ... 118
Gupta, Ram ... 122
Haeri, Foad ... 103
Hasse, Hans ... 47
Hawa, Hani ... 77
Haynes, Daniel ... 108
Heider, Patrick ... 12
Hernandez, Sergio ... 122
Hesketh, Robert ... 9, 14
Hiaki, Toshihiko ... 96
Hirohama, Seiya ... 100
Hoepfner, Michael P. ... 97
Hu, Jianli ... 75, 106, 110, 114, 140
Hu, Sheng ... 101, 105
Huang, Yu-Chao ... 137
Hussain, Syed ... 86
Huynh, Khang ... 122
Hwang, Hyun Tae ... 87
Ilsa, Kristiina ... 2
Ince, Nevin Gerek ... 100
Islam, Md Tahmid ... 120, 123
Iv, Thomas Knotts ... 98
Jacobi, David ... 68
Jacquemond, Rémy ... 66
Jahromi, Hossein ... 124
Jang, Joonbaek ... 137
Jangam, Kalyani ... 80
Jeffryes, Clayton ... 17
Jerpoth, Swapana ... 9, 14
Jessen, Kristian ... 101, 105
Jiang, Changle ... 75, 114
Jiao, Feng ... 63
Jolaoso, Lateef ... 81
Joshi, Anuj ... 80, 111
Joshi, Pankti ... 17
Junaedi, Christian ... 77
Jupke, Andreas ... 35
Kamal, Muhammad Shahzad ... 86
Karimi, Iftekhar ... 69
Karpe, Sanjana ... 106
Karunarathne, Sampath ... 4, 60
Kaur, Manjodh ... 33
Kazempoor, Pejman ... 81
Kim, Jinsoo ... 87, 88
Kim, Ki-Joong ... 58
Kim, Seung-Soo ... 87, 88
Kline, Matthew J. ... 4, 60
Klinger, Jordan ... 120, 123
Knoetze, Johannes ... 90
Kong, Fanhe ... 74
Kong, Weiyi ... 97
Kumar, Anand ... 78
Kumar, Sandeep ... 122
Kumar, Sonu ... 111
Kumari, Geeta ... 121
Kwon, Joseph ... 89
Lao, Jianpei ... 118
Lapkin, Alexei A. ... 34
Leitl, Matthias ... 65
Lemaire III, Peter ... 103
Li, Fanxing ... 113
Li, Wenjing ... 5
Lin, Yu-Ting ... 89
Liu, Junli ... 126
Liu, Nian ... 30
Liu, Shuhao ... 89
Lott, Patrick ... 130
Loughney, Patricia ... 111
Lu, Mi ... 118
Lubenau, Udo ... 47
Luján, Andrea Loya ... 8
Luo, Hongmei ... 62
Ly, Hoang Vu ... 87
Maddipudi, Bharathkiran ... 122
Maestri, Matteo ... 129
Mahboob, Ahmad ... 86
Maheshwari, Utkarsh ... 93
Marca, Josep ... 50
Marsden, Grant ... 72
Masnadi, Mohammad S. ... 139
Matranga, Christopher ... 58
McGaughy, Kayle ... 139
McKone, James R. ... 33
Mendes, Pedro ... 133
Meng, Qichang ... 111
Meramo, Samir ... 115
Mesdour, Rabah ... 71

Mhatre, Dwijraj....28
Mielke, Kristina....35
Mitsos, Alexander....65
Miu, Evan....33
Moellenhoff, Caleb....127
Mohammad, Zain....80
Mohapatra, Pinak....74, 111
Morales-Guio, Carlos....137
Muhich, Christopher....11
Mukarakate, Calvin....2
Mukasyan, Alexander....78
Muley, Pranjali....108
Munoz, Carlos....37
Murtaza, Mobeen....86
Nair, Sankar....118
Neal, Luke....113
Norton Jr., John....7
Oba, Shigeo....96
Ocampo, Kaylen....139
Overa, Sean....63
Paffenroth, Randy....5
Pal, Ajinkya....69
Panja, Palash....104
Pant, Kamal....28, 76
Paraye, Tanashree....93
Parmar, Kaushal....76
Patel, Kanhaiyalal....71
Patil, Shirish....86
Patra, Debshilpi....84
Paudel, Hari....108
Paudel, Shyam....127
Pawar, Prasad....17
Pendse, Hemant P....4
Peres, Jose....18
Perkins, David....102
Pham, Hien N....129
Phillips, Ailene G....12
Prokopou, Georgia Ioanna....65
Qin, Lang....111
Qubian, Ali....18
Rahman, Rizwanur....97
Ramani, Vijay....31
Ramasamy, Karthikeyan....1, 7
Ramteke, Akshata....28
Realff, Matthew....118
Reyes, Victoria....79
Reza, Toufiq....120, 123
Riazi, M R....18
Rimer, Jeffrey....129
Rivera, Daniel J....11
Robinson, Brandon....75, 106, 110, 140
Rodriguez, Jennifer....68
Roenigk, Samantha....33
Rogers, William....58
Rowane, Aaron....99
Roy, Shantanu....76
Roy, Shyamal....117
Roychoudhury, Subir....77
Ruan, Chongyan....113
Saathoff, Jonathan D....102
Saha, Eshita....117
Saha, Nepu....120
Sanguinito, Sean....103
Sankarasubramanian, Shrihari....31
Santos, Vera....12
Santosa, Daniel....2, 7
Sanyal, Udishnu....1
Savelski, Mariano J....9, 14
Schaidle, Joshua....2
Schardt, Sven....130
Schmidt, Andrew J....7
Schuman, Thomas....125
Schunk, Stephan A....130
Schwartz, Thomas....4
Schwarz, Cara....90
Seidenberg, Jan Raphael....65
Shabuddin, Muntasir....5
Shafirovich, Evgeny....79
Shah, Parth....103
Shahnam, Mehrdad....58
Sharma, Kritika....31
Shaw, Alexander....72
Shell, Katelyn....122
Shen, Kangze....137
Shende, Anuradha....122
Shende, Rajesh....122
Shishkova, I....18
Slater, C. Stewart....9, 14
Sloley, Andrew....26, 94
Smith, Joseph....59, 127
Smith, Stuart E....102
Song, Tao....125
Soni, Bhavin....121
Soriano Jr., Nestor....50
Sorkhabi, Rasoul....104
Srinivasan, Poorna....68
Stangland, Eric....12
Stefan, Van Bloois....12
Stratiev, D....18
Stringari, Paolo....100
Stuart, Leadley....12
Subramaniam, Senthil....1, 7
Sukumara, Sumesh....115
Sum, Amadeu....73

Sunny, Ashin 111
Susman, Mariano D. 129
Tao, Bernard 126
Tapriyal, Deepak 103
Thakkar, Anuj 122
Thies, Mark C. 102
Thorson, Michael R. 7
Thybaut, Joris 133
Timko, Michael T. 5
Tindall, Graham W. 102
Tocha, Ewa 12
Tong, Andrew 74
Tsotsis, Theodore 101, 105
Twizerimana, Aime Laurent 139
Urban-Klaehn, Jagoda M. 62
Vaidya, Ravi 71
Van Ommen, J Ruud 64
Van Steijn, Volkert 64
Vera, Clarita Regalado 62
Veser, Goetz 106, 139
Vilekar, Saurabh A. 77
Voggenreiter, Johannes 47
Wan, Charles 66
Wang, Huamin 2
Wang, Xiaoyan 140
Wang, Yuxin 140
Wang, Zhongyang 31
West, David 129
Wheeler, Clayton 4
Wheeler, M. Clayton 60
Wilding, Wade Vincent 98
Wolf, Eduardo E. 78
Worch, Denis 47
Wu, Jiyue 105
Wu, Ning 73
Wu, Yue 128
Xu, Gang 100
Yang, Jin 12
Yenkie, Kirti 9, 14
Zenasni, Oussama 50
Zhang, Yitao 74
Zhang, Yuanxing 73
Zhang, Yunlong 102
Zhao, Yang 85
Zhou, Meng 62